The *Mathematics of Investment*
Mathematics for Everyday Living

Roland E. Larson
Robert P. Hostetler

Exercises prepared with the assistance of
David E. Heyd

Updated and revised by
Marjorie J. Bertram
Richard Bertram

This book is published by
Meridian Creative Group, a Division of Larson Texts, Inc.

(800) 530-2355
http://www.meridiancg.com

Trademark acknowledgement:
Explorer Plus is a trademark of Texas Instruments Incorporated

Copyright © 1997 by Meridian Creative Group

All rights reserved. No part of this publication may be reproduced or transmitted in any form or by any means, electronic or mechanical, including photocopying, recording, or any information storage or retrieval system, without prior permission from the publisher.

Printed in the United States of America

International Standard Book Number: ISBN 1-887050-29-9

10 9 8 7 6 5 4

Preface

Mathematics for Everyday Living is a series of workbooks designed to give students a solid and practical grasp of the mathematics used in daily life.

Each workbook in the series features practical consumer information, solved examples, and "Try one!" exercises with which students can use a particular skill immediately following its introduction. Complete solutions to the "Try one!" exercises are provided in the back of the workbook along with answers to odd-numbered exercises.

The workbook, *"The Mathematics of Investment,"* is divided into five sections, each of which address a set of mathematical skills and concepts involved in the investment of money.

- The first section, "Common Stocks: An Introduction," shows how to calculate stock dividends, dividend yields, and price-earning ratios. The section also describes different classifications of stocks.

- Section two, "Common Stocks: Buying and Selling," shows students how to read and interpret the financial section of major newspapers and discusses the calculations involved in the purchase and sale of stocks.

- In the third section, "Corporate Bonds," the student will learn how to read and interpret the bond report in most major newpapers as well as how to calculate the interest, purchase price, and selling price of bonds.

- In section four, "Mutual Funds," the student will be introduced to the terminology and calculations necessary to understand mutual fund transactions.

- Finally, section five features a spotlight on a career in finance. Educational requirements, job descriptions and employment outlook are all addressed.

Instructors may download a Teacher's Resource Guide for this title on our website (http://home.meridiancg.com/mfelpdf.html). The calculator keystrokes in this workbook are provided for the Texas Instruments Explorer Plus™.

Finally, thank you to Dr. Timothy Smaby, Assistant Professor of Finance at The Pennsylvania State University, Behrend College, for his technical assistance and guidance, and to Cheryl Bernik for her help in organizing this project.

Marjorie J. Bertram

Contents

Section 1   Common Stocks: An Introduction   1

Section 2   Common Stocks: Buying and Selling   33

Section 3   Corporate Bonds   59

Section 4   Mutual Funds   87

Section 5   Spotlight on the Financial Manager   105

Solutions to "Try one!" Exercises   115
Answers to Odd-Numbered Exercises   125

Section 1
Common Stocks: An Introduction

Business ownership in the United States falls into three basic categories: **sole proprietorships, partnerships,** and **corporations**. Almost all large businesses are in the third category and as such they have a distinct legal existence that is separate from the person or persons owning the business. (Under normal conditions, the owners of a corporation are not personally liable for corporation debts, while owners of sole proprietorships and partnerships are liable for debts incurred by their businesses.)

When a business is incorporated, a certain number of **shares** of the company's stock are issued. These shares may be owned by one or more **stockholders**. In the case of large corporations such as Chrysler or IBM, there may be thousands of stockholders. (In 1996, Chrysler had 136,100 stockholders and IBM had 652,923 stockholders.) Although stockholders are the owners of a corporation, they are not usually involved in the day-to-day operations of the corporation. Normally, a board of directors is elected to manage the corporation.

As with all businesses, corporations exist for the purpose of earning a profit. Since a corporation is a separate legal entity, corporate earnings may be retained by the corporation or returned to the stockholders (usually on a quarterly basis) in the form of dividends. The decision to retain earnings or pay dividends is rarely made on an "all or nothing" basis. Typically, the board of directors of the corporation decides to return a percent of the earnings to stockholders and retain the remainder for corporate growth. Sometimes the corporation will not pay dividends at all.

There are two basic types of corporate stocks: **preferred stock** and **common stock**. Dividends on preferred stocks are paid at a fixed percentage of **par value** (an artificial value assigned at the time of issue) and they are given a preference on dividends. That is, available dividends must be paid to holders of preferred stocks before any are paid to holders of common stock. From an investment point of view, preferred stocks are roughly comparable to corporate bonds (see Section 3). In this and the following section, our discussion of stocks is limited to common stocks.

Dividends on common stocks are paid as follows:

$$d = \frac{D}{N}$$

where

d = annual dividend per share

D = total annual dividends available to common stockholders

N = total number of shares of common stock outstanding.

EXAMPLE 1 Finding the dividend per share of common stock

Marlene Livingstone owns 300 shares of common stock in a utility company. In 1995 the company paid a total dividend of $479,400,000 to the owners of its 235 million shares of common stock. How much did Marlene receive in dividends for her 300 shares in 1995? If the dividends were paid in four equal payments, how much did Marlene receive each quarter?

SOLUTION

To find the dividend per share for 1995, we use $N = 235,000,000$ and $D = \$479,400,000$. Therefore, the dividend per share is

$$d = \frac{D}{N} = \frac{\$479,400,000}{235,000,000} = \$2.04.$$

Since Marlene owns 300 shares of this common stock, her annual dividend AD is

$$AD = (300)(\$2.04) = \$612.00.$$

Therefore, each quarter Marlene receives

$$\text{Quarterly Dividend} = \frac{AD}{4} = \frac{\$612.00}{4} = \$153.00. \quad \blacklozenge$$

COMMON STOCKS: AN INTRODUCTION

Try one!

Robert Cunningham owns 200 shares of common stock in a publishing company. In 1995 the company paid a total dividend of $190,350,000 to the owners of its 141 million shares of common stock. How much did Robert receive in dividends for his 200 shares in 1995? If the dividends were paid in four equal payments, how much did Robert receive each quarter?

Total 1995 Dividend: _____   Quarterly Payment: _____

Most stocks are purchased through brokerage firms and there are no fixed prices for stocks. In fact, the prices of most stocks fluctuate from day to day. To buy shares of stock on an organized exchange, an investor can place an order with a brokerage firm. The broker will give the most recent price of the desired stock and if that is agreeable to the investor, the order is sent electronically to the firm's partner on the floor of a stock exchange. If a seller is found at that asking price, the sale is negotiated and then reported to the brokerage firm.

If you decide to invest in stocks, you must decide upon one or more corporations that are financially sound enough to give a good return on your investment. A factor in assessing the financial well-being of a corporation is its current **dividend yield**, which is the per share ratio of the annual dividend to the price of the stock.

$$\text{Dividend Yield} = \frac{\text{Annual Dividend Per Share}}{\text{Price Per Share}} = \frac{d}{p}$$

Table 1 lists the dividend yield based on the average price of four common stocks over the years from 1986 to 1995. Note that the **price per share p** is given in dollars and eighths of dollars. Share prices can also be reported in sixteenths or even thirty-seconds of a dollar.

TABLE 1

Selected Common Stock Performances from 1986 to 1995*

| | General Motors ||| Gannett ||| Nike ||| Delta Airlines |||
	Dividend	Aver. Price	Current %Yield	Dividend	Aver. Price	Current %Yield	Dividend	Aver. Price	Current %Yield	Dividend	Aver. Price	Current %Yield
1986	$2.50	$37\frac{3}{8}$	6.7%	$0.84	$36\frac{1}{2}$	2.3%	$0.08	$2\frac{7}{8}$	2.8%	$1.00	$45\frac{1}{2}$	2.2%
1987	$2.50	$36\frac{1}{4}$	6.9%	$0.92	$41\frac{7}{8}$	2.2%	$0.10	$3\frac{7}{8}$	2.6%	$1.10	50	2.2%
1988	$2.50	$36\frac{3}{4}$	6.8%	$1.00	$34\frac{1}{2}$	2.9%	$0.10	$5\frac{7}{8}$	1.7%	$1.20	$57\frac{1}{8}$	2.1%
1989	$2.50	$44\frac{5}{8}$	5.6%	$1.08	$41\frac{1}{2}$	2.6%	$0.15	$11\frac{1}{2}$	1.3%	$1.70	68	2.5%
1990	$3.00	$41\frac{5}{8}$	7.2%	$1.20	$37\frac{1}{2}$	3.2%	$0.10	$9\frac{1}{8}$	1.1%	$1.20	$66\frac{5}{8}$	1.8%
1991	$1.60	$35\frac{1}{2}$	4.5%	$1.24	$41\frac{3}{8}$	3.0%	$0.28	28	1.0%	$1.20	$66\frac{5}{8}$	1.8%
1992	$1.40	$36\frac{7}{8}$	3.8%	$1.25	$48\frac{1}{8}$	2.6%	$0.30	$37\frac{1}{2}$	0.8%	$1.20	60	2.0%
1993	$0.80	$44\frac{1}{2}$	1.8%	$1.29	$51\frac{5}{8}$	2.5%	$0.40	$33\frac{3}{8}$	1.2%	$0.20	50	0.4%
1994	$0.80	50	1.6%	$1.33	$53\frac{1}{4}$	2.5%	$0.40	$30\frac{3}{4}$	1.3%	$0.20	50	0.4%
1995	$1.10	$45\frac{7}{8}$	2.4%	$1.37	$57\frac{1}{8}$	2.4%	$0.50	$55\frac{1}{2}$	0.9%	$0.20	$66\frac{5}{8}$	0.3%

*All stock prices except for Delta Airlines are adjusted for stock splits (explained in Section 2).

When interpreting Table 1, it is important to note that the dividend yield for a given year is valid *only* if the stock was purchased at the average price that year. If purchased at a different price, the dividend yield can be significantly higher or lower.

EXAMPLE 2   Finding the current dividend yield

Suppose Marlene Livingstone's 300 shares (see Example 1) had an average price of $43\frac{3}{8}$ per share in 1995. What was the dividend yield based upon the average price per share for Marlene Livingstone's utility stock in 1995?

COMMON STOCKS: AN INTRODUCTION   SECTION 1 **5**

SOLUTION

Since Marlene's 1995 dividend per share was $2.04 and the price per share at that time was $43\frac{3}{8}$, the dividend yield was

$$\text{Dividend Yield} = \frac{d}{p} = \frac{\$2.04}{43\frac{3}{8}} = \frac{\$2.04}{\$43.375} \approx 0.047 = 4.7\%. \blacklozenge$$

Note in Example 2 that $43\frac{3}{8}$ was changed to the decimal form 43.375 before dividing. The decimal forms for the various multiples of one-eighth are:

$\frac{1}{8}$	$\frac{2}{8} = \frac{1}{4}$	$\frac{3}{8}$	$\frac{4}{8} = \frac{1}{2}$	$\frac{5}{8}$	$\frac{6}{8} = \frac{3}{4}$	$\frac{7}{8}$
0.125	0.25	0.375	0.5	0.625	0.75	0.875

Calculator Hints

To find the decimal equivalent of a fraction using your calculator, apply the following steps:

1. Enter the numerator (top number).

2. Press the ÷ key.

3. Enter the denominator (bottom number).

4. Press =.

Try one!

Suppose that the average price per share for Robert Cunningham's investment (see the previous Try One!) was $50\frac{5}{8}$. What was the dividend yield based upon the average price per share for Robert's investment in 1995?

Answer: _____

A second and more significant indication of a corporation's well-being is its **earnings per share**. To find the earnings per share, we divide the total corporate earnings available to common stock by the number of shares of common stock.

$$\text{Earnings Per Share} = e = \frac{E}{N}$$

where

E = total corporate earnings available to common stock

N = number of shares of common stock outstanding.

EXAMPLE 3 Finding the earnings per share

Find the earnings per share for the following two corporations.

a. Johnson and Johnson:
1995 Earnings.......................... $2,409,369,600
Number of Shares................... 1,295,360,000

b. Chrysler Corporation:
1995 Earnings.......................... $2,004,990,000
Number of Shares................... 756,600,000

SOLUTION

a. The earnings per share for Johnson and Johnson for 1995 were

$$e = \frac{E}{N} = \frac{\$2,409,369,600}{1,295,360,000} = \$1.86.$$

b. The earnings per share for Chrysler Corporation for 1995 were

$$e = \frac{E}{N} = \frac{\$2,004,990,000}{756,600,000} = \$2.65.$$

COMMON STOCKS: AN INTRODUCTION   SECTION 1   **7**

Try one!

Find the earnings per share for the following corporation.

Lockheed Martin Corporation:
1995 Earnings.. $124,539,480
Number of Shares..   73,692,000

Answer: _____

A third measure of a corporation's well-being is the **price-earnings ratio** (abbreviated *P-E* ratio). The price-earnings ratio is the ratio of the price per share to the corporation's annual earnings per share.

$$\text{Price-Earnings Ratio} = \frac{\text{Price Per Share}}{\text{Earnings Per Share}} = \frac{p}{e}$$

EXAMPLE 4   **Finding the price-earnings ratio**

For 1995, the board of directors of Proctor & Gamble reported the following corporate earnings available to common stocks:

    1995 Earnings..................................... $2,547,189,540
    Number of Shares..............................   686,574,000

The average price per share was 79 in 1995. What was the price-earnings ratio?

SOLUTION

The earnings per share for 1995 were

$$e = \frac{E}{N} = \frac{\$2{,}547{,}189{,}540}{686{,}574{,}000} = \$3.71.$$

Therefore, the price-earnings ratio for a price of 79 was

$$\text{Price-Earnings Ratio} = \frac{p}{e} = \frac{\$79.00}{\$3.71} \approx 21.3. \qquad \blacklozenge$$

Try one!

For 1995, the board of directors of Lockheed Martin Corporation reported the following corporate earnings available to common stocks.

1995 Earnings............................	$651,414,560
Number of Shares...................	198,602,000

The average price per share was $65\frac{5}{8}$ in 1995. What was the price-earnings ratio?

Answer: _____

Stock prices are extremely temperamental and often show a greater correlation with the general state of the economy than the actual performance record of a corporation. For example, during the ten-year period from 1986 to 1995, the market prices of Hewlett-Packard Company had a relatively steady earning and dividend growth throughout the entire decade. (See Figures 1 and 2 and Table 2.)

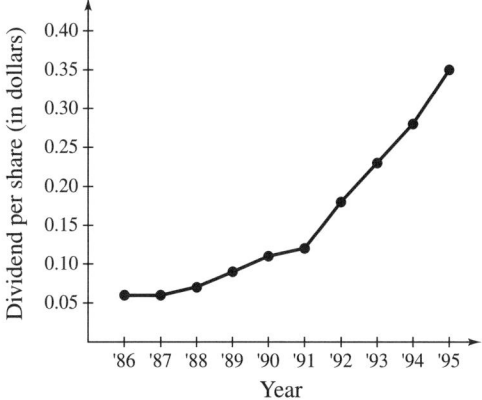

Figure 1
Hewlett-Packard Dividends (1986-1995)

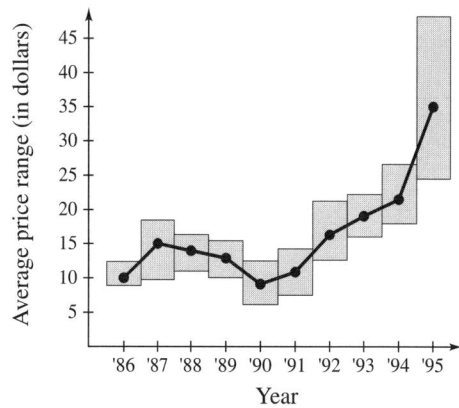

Figure 2
Hewlett-Packard Prices (1986-1995)

TABLE 2

Hewlett-Packard Company Common Stock (1986 to 1995)

Year	Earnings Per Share	Dividends Per Share	Dividend Yield	Price Range Low	Price Range Average	Price Range High	Price Earnings Ratio
1986	$0.51	$0.06	0.5%–0.7%	$8\frac{7}{8}$	10	$12\frac{3}{8}$	17.4–24.3
1987	$0.63	$0.06	0.3%–0.6%	$9\frac{3}{4}$	15	$18\frac{1}{2}$	15.5–29.4
1988	$0.84	$0.07	0.4%–0.6%	11	14	$16\frac{3}{8}$	13.1–19.5
1989	$0.88	$0.09	0.6%–0.9%	10	$12\frac{7}{8}$	$15\frac{3}{8}$	11.4–17.5
1990	$0.77	$0.11	0.9%–1.8%	$6\frac{1}{8}$	$9\frac{1}{8}$	$12\frac{1}{2}$	8.0–16.2
1991	$0.76	$0.12	0.8%–1.6%	$7\frac{1}{2}$	$10\frac{7}{8}$	$14\frac{1}{4}$	9.9–18.8
1992	$0.88	$0.18	0.8%–1.4%	$12\frac{5}{8}$	$16\frac{3}{8}$	$21\frac{1}{4}$	14.4–24.1
1993	$1.16	$0.23	1.0%–1.4%	16	$19\frac{1}{8}$	$22\frac{1}{4}$	13.8–19.2
1994	$1.54	$0.28	1.0%–1.6%	18	$21\frac{1}{2}$	$26\frac{5}{8}$	11.7–17.3
1995	$2.32	$0.35	0.7%–1.4%	$24\frac{1}{2}$	35	$48\frac{1}{4}$	10.6–20.8

The most often quoted statistic showing the trend in stock prices is the *Dow–Jones Industrial Average.* This average is derived from the current stock prices of 30 selected large corporations that are chosen on the basis of their representation of the United States economy. Figures 3 and 4 compare the Dow–Jones Industrial Averages and the Economic Growth Rate for the United States during the period from 1986 to 1995. (The **Economic Growth Rate** is the change in the *Gross Domestic Product* adjusted for inflation.) Note that when the Economic Growth Rate falls below zero, the country is said to be in a state of *recession.*

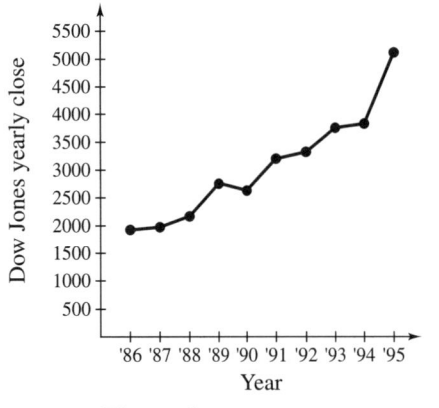

Figure 3
Stock Prices

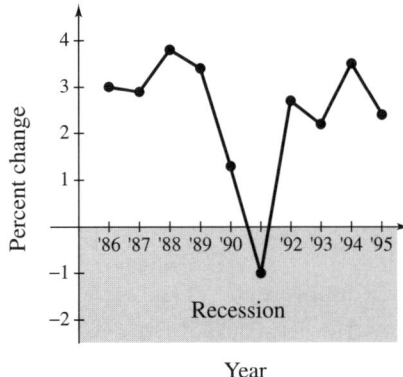

Figure 4
U.S. Economic Growth Rate

In attempting to predict future performances of stocks, investors often classify stocks in categories such as cyclical stocks and defensive stocks, or growth stocks and value stocks.

Cyclical stocks are characterized by dividend changes that tend to be more sensitive to the general state of the economy. (Note that most stock *prices* change with the general state of the economy, but with cyclical stocks, the *dividends* also tend to follow this pattern of change.) Historically, General Motors stock has been cyclical in that the changes in General Motors common stock dividends roughly correlate with the Economic Growth Rate. To see this, compare Figure 4 with Figure 5.

Figure 5
General Motors Dividends
(1986–1995)

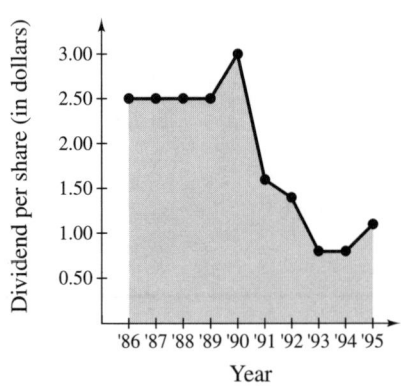

When looking back upon stock investments, cyclical stock turns out to be a good investment during times of recession, especially if it happens to be purchased at a low price near the end of a recessionary period. The following example illustrates a situation in which an individual happened to purchase a cyclical stock at an ideal time.

EXAMPLE 5   Finding the dividend yield

Bruce Dover purchased 200 shares of General Motors common stock on January 2, 1991 at $34\frac{1}{4}$ per share. The closing prices for 1991, 1992, and 1993 are as follows.

Year	Closing Price
1991	$28\frac{7}{8}$
1992	$32\frac{1}{4}$
1993	$54\frac{7}{8}$

a. Use Table 1 to find Bruce's total dividends during 1991, 1992, and 1993.

b. What was the dividend yield (based upon the closing price) for Bruce's investment during each of these years?

SOLUTION

a. From Table 1, we see that General Motors paid dividends of $1.60, $1.40, and $0.80 per share during 1991, 1992, and 1993, respectively. Since Bruce owns 200 shares, his total dividends were

1991 dividends = (200)($1.60) = $320.00

1992 dividends = (200)($1.40) = $280.00

1993 dividends = (200)($0.80) = $160.00.

b. The dividend yield for each of these years was

$$1991 \text{ dividend yield} = \frac{\$1.60}{\$28.875} \approx 0.055 = 5.5\%$$

$$1992 \text{ dividend yield} = \frac{\$1.40}{\$32.25} \approx 0.043 = 4.3\%$$

$$1993 \text{ dividend yield} = \frac{\$0.80}{\$54.875} \approx 0.015 = 1.5\%.$$ ◆

Try one!

a. Suppose Steve Johnson purchased 300 shares of General Motors stock in 1990 at $39\frac{1}{4}$ per share. Using Table 1, find Steve's total dividends during 1993, 1994, and 1995.

Total Dividends 1993: _____

Total Dividends 1994: _____

Total Dividends 1995: _____

b. Determine the dividend yield on October 24, 1995 if the closing price on that date was $44\frac{5}{8}$.

Dividend Yield: _____

Defensive stocks are characterized by a dividend and earnings record that is less affected by the general state of the economy. Whether business is good or bad on a national scale, defensive stocks tend to go right on paying dividends at a fixed or slightly increasing rate. Utility company stocks are usually defensive as are stocks of other corporations dealing with day-to-day necessities such as pharmaceuticals and staple food products.

The major advantage of defensive stocks (a predictable dividend rate) is often a disadvantage during inflationary times. The reason is that during inflationary times, *risk-free* investments such as savings certificates and treasury bonds often yield more than defensive (low-risk) stocks. The result is that defensive stocks tend to drop in price during periods of high inflation and high interest rates. (Note: *all* stocks tend to drop in price during periods of high inflation; however, defensive stock tend to show a greater drop in price.) This is demonstrated graphically in Figure 6, which compares the average utility common stock price from 1976 to 1994 with the annual rate of inflation during that same period.

Figure 6
Utility Common Stocks and Inflation (1976-1994)

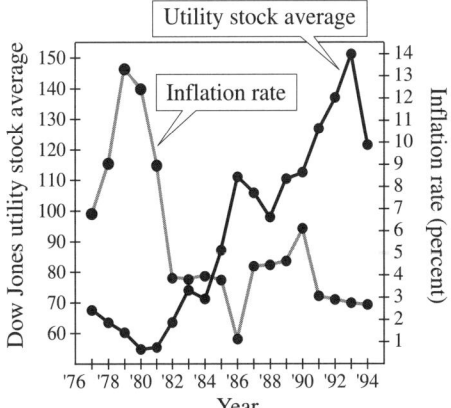

As long as an investor keeps a particular stock, the current dividend yield is a measure similar to the annual percentage rate for savings accounts. However, when an investor sells stock, a different measure comes into play. This measure is called the **total income** of a stock.

Total Income = Selling Price − Purchase Price + Dividends

When stock prices drop, the loss in selling price can easily consume any dividends which the investor received. The next example illustrates the disastrous effect that stock prices can have on the total income.

EXAMPLE 6 — Finding the total income on common stocks

Arlene Thompson purchased 1000 shares of Eastern Utilities Associates common stock at 38 per share in 1986. In 1995 Arlene decided to take her losses and she sold at $23\frac{3}{8}$. Use the following table to determine Arlene's total income.

Year	1986	1987	1988	1989	1990	1991	1992	1993	1994	1995
Dividend	$2.15	$2.27	$2.38	$2.48	$2.58	$1.45	$1.36	$1.42	$1.52	$1.59

(Assume Arlene collected the full dividends during each of the ten years, and ignore broker's fees.)

SOLUTION

Arlene's dividends per share over the ten years totaled

$$\text{Dividends Per Share} = \$2.15 + \$2.27 + \$2.38 + \$2.48 + \$2.58$$
$$+ \$1.45 + \$1.36 + \$1.42 + \$1.52 + \$1.59 = \$19.20.$$

Since Arlene had 1000 shares, her total dividends during the ten-year period were

$$\text{Dividends} = (1000)(\$19.20) = \$19{,}200.00.$$

Since Arlene paid 38 per share and sold at $23\frac{3}{8}$, we have

$$\text{Purchase Price} = (1000)(\$38.00) = \$38{,}000.00$$

$$\text{Selling Price} = (1000)(\$23.375) = \$23{,}375.00.$$

Therefore, Arlene's total income was

$$\text{Total Income} = \$23{,}375.00 - \$38{,}000.00 + \$19{,}200.00$$

$$= \$4{,}575.00. \quad \blacklozenge$$

COMMON STOCKS: AN INTRODUCTION   SECTION 1   **15**

Try one!

Suppose that Arlene Thompson's sister Madeline (see Example 6) bought 500 shares of Eastern Utilities Associates common stock in 1986 at 38 but, unlike Arlene, she decided to sell in 1991 at 17. Determine Madeline's total income. Assume Madeline collected the full dividends for each of the years she held the stock.

Answer: _____

Keep in mind that the total income in Example 6 represents ten years' worth of investments. Very roughly, this total income represents a 1.2% rate of return per year. (This is a poorly performing investment. Most banks have savings accounts available with interest rates better than the approximate return on this particular investment.)

One way to measure an investment's performance is the **annual return**

$$\text{Annual Return} = \frac{\text{Ending Price} - \text{Beginning Price} + \text{Dividends}}{\text{Beginning Price}}$$

where the ending price is the price of the stock at the end of the year and the beginning price is the price of the stock at the beginning of the year.

EXAMPLE 7 **Finding the annual return on common stock**

Collin Spears invested $1187.50 in Nike common stock at $11\frac{7}{8}$ per share on January 3, 1994. On December 31, 1994, the closing price for Nike stock was $18\frac{5}{8}$ and Collin wanted to know how well his investment was doing so that he could compare it to his bank account's performance. Determine Collin's annual dividend and annual return for this investment.

SOLUTION

At $11.875 per share, Collin's investment of $1187.50 would buy

$$\frac{\$1187.50}{\$11.875} = 100 \text{ shares}$$

of Nike common stock. From Table 1 we know that Collin's dividend per share for 1994 was $0.40.

Since Collin had 100 shares, his dividends during the year were

Annual Dividends = (100)($0.40) = $40.00.

Since Collin paid $11.875 per share and the closing price for the year was $18.625, we have

Beginning price = $1187.50

Ending price = (100)($18.625) = $1862.50.

Thus, Collin's annual return was

$$\text{Annual Return} = \frac{\text{Ending Price} - \text{Beginning Price} + \text{Dividends}}{\text{Beginning Price}}$$

$$= \frac{\$1862.50 - \$1187.50 + \$40.00}{\$1187.50}$$

$$= \frac{\$715.00}{\$1187.50} \approx 0.602 = 60.2\%.$$ ◆

The reason Collin did so well on his Nike stock in Example 7 is that he purchased the stock at a time when few investors believed it would become a **growth stock**.

Growth stocks are characterized by a high annual growth rate of earnings. Nike (see Table 1) was considered a growth stock during the late 1980s and early 1990s. Other characteristics of growth stocks are their low dividend payout percentages (the corporation retains most of its earnings for future growth) and high price-earnings ratios (investors expect the price per share to increase).

The last type of common stock is **value stock**. We reserve this term for those stocks that have *poor* past records (indicated by a low price-earnings ratio). No one wants a stock whose dividends and prices *continue* to fall. Yet many investors are willing to buy stocks whose dividends and prices have fallen because they speculate that this pattern will reverse. Value stock purchases have occasionally paid off in dramatic proportions. For example, IBM (International Business Machines Corporation) common stock rose in value from $59\frac{7}{8}$ per share in 1993 to $114\frac{5}{8}$ per share in 1995. Earlier, this same stock dropped from $161\frac{7}{8}$ in 1986 to $48\frac{3}{4}$ in 1992. See Figure 7 to observe the ups and downs of IBM stock prices.

Figure 7
IBM Common Stock Prices (1986-1995). (Prices adjusted for splits.)

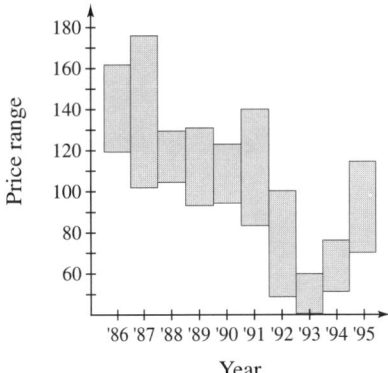

Our last Try One! describes a "win" with IBM common stock.

Try one!

George Walters had been following the price of IBM common stock. In 1993 he felt that the stock was as low as it would go and withdrew $4062.50 from his savings account to purchase 100 shares at $40\frac{5}{8}$. One year later, George sold his shares at $62\frac{3}{8}$. What was his annual return? (Assume that George received dividends of $1.00 per share during the time he held the stock.)

Answer: _____

Important Terms

annual return	N, number of shares
common stock	p, price per share
corporation	partnership
cyclical stock	par value
d, annual dividend per share	preferred stock
D, total annual dividend available to common stockholders	price-earnings ratio
defensive stock	shares
dividend yield	sole proprietorship
e, earnings per share	stockholders
E, total earnings	total income
economic growth rate	value stock
growth stock	

Important Formulas

$$\text{Annual Return} = \frac{\text{Ending Price} - \text{Beginning Price} + \text{Dividends}}{\text{Beginning Price}}$$

$$d = \frac{D}{N}$$

where

d = annual dividend per share

D = total annual dividends available to common stockholders

N = total number of shares of common stock outstanding.

$$\text{Dividend Yield} = \frac{\text{Annual Dividend Per Share}}{\text{Price Per Share}} = \frac{d}{p}$$

$$\text{Earnings Per Share} = e = \frac{E}{N} = \frac{\text{total earnings}}{\text{number of shares}}$$

$$\text{Price–Earnings Ratio} = \frac{\text{Price Per Share}}{\text{Earnings Per Share}} = \frac{p}{e}$$

Total Income = Selling Price − Purchase Price + Dividends

CONSUMER HINTS

- Stock market investments can pay handsome returns but they should represent only a portion of your total investment plan. While potential rewards are great, a risk factor is also present.

- Be wary of investment advice from unfamiliar sources. In the stock market, there is no such thing as a "sure thing."

- Keep track of your own stock on a daily (or at least weekly) basis. Most newspapers carry a listing of current stock prices and trends, and there are numerous internet sites on the world wide web that enable you to track your own portfolio.

- If you own several stocks, try to "diversify" by owning stocks in several different industries. Do not put all your eggs in one basket.

COMMON STOCKS: AN INTRODUCTION   SECTION 1   **21**

SECTION 1   EXERCISES

In Exercises 1–6, determine (a) the annual dividend per share and (b) the dividend yield.

1. A corporation paying a total common stock dividend of $93,680,000 on 47,661,000 shares selling for $45.00 per share.

   Answer (a) _____

   Answer (b) _____

2. A corporation paying a total common stock dividend of $8,315,720 on 8,484,000 shares selling for $21.00 per share.

   Answer (a) _____

   Answer (b) _____

22 SECTION 1   COMMON STOCKS: AN INTRODUCTION

3. A corporation paying a total common stock dividend of $30,346,360 on 14,199,000 shares selling at $29\frac{3}{8}$.

Answer (a) _____

Answer (b) _____

4. A corporation declares 22% of its net income of $55,799,000 for common stock dividends on 20,976,000 shares selling at 24.

Answer (a) _____

Answer (b) _____

5. A corporation declares 20% of its net corporate earnings of $11,550,000 for common stock dividends on 4,534,000 shares selling at $54\frac{1}{8}$.

Answer (a) _____

Answer (b) _____

6. A corporation declares 71% of its net corporate earnings of $61,236,000 for common stock dividends on 32,747,000 shares selling at $12\frac{5}{8}$.

Answer (a) _____

Answer (b) _____

In Exercises 7–10, complete the table determining the earnings per share and the price-earnings ratio.

7. Sears, Roebuck and Company

Year	Earnings	Number of Shares	Average Price per Share	Earnings per Share	Price-Earnings Ratio
1991	$1,276,611,000	344,100,000	$33\frac{7}{8}$	_____	_____
1992	$2,427,516,000	345,800,000	$42\frac{1}{2}$	_____	_____
1993	$2,181,976,000	350,800,000	46	_____	_____
1994	$1,111,372,000	351,700,000	$48\frac{1}{2}$	_____	_____
1995	$987,965,000	390,500,000	$31\frac{1}{8}$	_____	_____

8. McDonald's Corporation

Year	Earnings	Number of Shares	Average Price per Share	Earnings per Share	Price-Earnings Ratio
1991	$846,296,000	717,200,000	$16\frac{1}{2}$	_____	_____
1992	$945,360,000	727,200,000	$21\frac{7}{8}$	_____	_____
1993	$1,032,804,000	707,400,000	$26\frac{1}{2}$	_____	_____
1994	$1,165,416,000	693,700,000	$29\frac{1}{4}$	_____	_____
1995	$1,378,409,000	699,700,000	$43\frac{7}{8}$	_____	_____

COMMON STOCKS: AN INTRODUCTION   SECTION 1   25

9. <u>Aetna, Incorporated</u>

Year	Earnings	Number of Shares	Average Price per Share	Earnings per Share	Price-Earnings Ratio
1991	$505,198,350	110,065,000	$40\frac{5}{8}$	_____	_____
1992	$5,513,500	110,270,000	$43\frac{3}{4}$	_____	_____
1993	$623,227,840	112,496,000	$55\frac{1}{4}$	_____	_____
1994	$466,404,120	112,658,000	$54\frac{1}{8}$	_____	_____
1995	$477,264,320	114,727,000	$62\frac{3}{4}$	_____	_____

10. <u>Mobil Corporation</u>

Year	Earnings	Number of Shares	Average Price per Share	Earnings per Share	Price-Earnings Ratio
1991	$1,852,099,650	398,301,000	$69\frac{1}{2}$	_____	_____
1992	$1,248,294,080	398,816,000	$65\frac{1}{4}$	_____	_____
1993	$2,018,711,760	398,168,000	65	_____	_____
1994	$1,694,824,360	395,987,000	$75\frac{1}{2}$	_____	_____
1995	$2,321,937,200	395,560,000	$84\frac{3}{8}$	_____	_____

In Exercises 11–16, find the dividend yield based upon the year-end closing price and the annual return on the stock for each year. (Assume that full dividends are collected for each year and ignore any broker's fees.) Note: The year-end closing price per share is the same as the beginning price per share for the following year.

11. Bought 100 shares of Caterpillar, Incorporated stock for 20 per share at the end of December, 1986.

Year	Annual Dividend per Share	Year-End Closing Price per Share	Dividend Yield Based Upon Closing Price	Annual Return
1987	$0.25	31	_____	_____
1988	$0.38	32	_____	_____
1989	$0.60	$28\frac{15}{16}$	_____	_____
1990	$0.60	$23\frac{1}{2}$	_____	_____
1991	$0.53	$21\frac{15}{16}$	_____	_____
1992	$0.30	$27\frac{1}{16}$	_____	_____
1993	$0.30	$44\frac{1}{2}$	_____	_____

12. Bought 100 shares of Aetna, Incorporated stock for $55\frac{1}{4}$ per share at the end of December, 1986.

Year	Annual Dividend per Share	Year-End Closing Price per Share	Dividend Yield Based Upon Closing Price	Annual Return
1987	$2.73	$45\frac{1}{4}$	_____	_____
1988	$2.76	$47\frac{1}{4}$	_____	_____
1989	$2.76	$56\frac{1}{2}$	_____	_____
1990	$2.76	39	_____	_____
1991	$2.76	44	_____	_____
1992	$2.76	$46\frac{1}{2}$	_____	_____
1993	$2.76	$60\frac{3}{8}$	_____	_____
1994	$2.76	$47\frac{1}{8}$	_____	_____
1995	$2.76	$69\frac{1}{4}$	_____	_____

28 SECTION 1   COMMON STOCKS: AN INTRODUCTION

13. Bought 500 shares of Hewlett-Packard Company stock for $10\frac{1}{8}$ per share at the end of December, 1986.

Year	Annual Dividend per Share	Year-End Closing Price per Share	Dividend Yield Based Upon Closing Price	Annual Return
1987	$0.06	$14\frac{9}{16}$	_____	_____
1988	$0.07	$13\frac{5}{16}$	_____	_____
1989	$0.09	$11\frac{11}{16}$	_____	_____
1990	$0.11	8	_____	_____
1991	$0.12	$14\frac{1}{4}$	_____	_____
1992	$0.18	$17\frac{1}{2}$	_____	_____
1993	$0.23	$19\frac{3}{4}$	_____	_____
1994	$0.28	$25\frac{5}{16}$	_____	_____

COMMON STOCKS: AN INTRODUCTION

14. Bought 500 shares of Schlumberger Limited stock for 32 per share at the end of December, 1986.

Year	Annual Dividend per Share	Year-End Closing Price per Share	Dividend Yield Based Upon Closing Price	Annual Return
1987	$1.20	$28\frac{3}{4}$	_____	_____
1988	$1.20	$32\frac{5}{8}$	_____	_____
1989	$1.20	$49\frac{1}{8}$	_____	_____
1990	$1.20	$57\frac{7}{8}$	_____	_____
1991	$1.20	$62\frac{3}{8}$	_____	_____
1992	$1.20	$57\frac{1}{4}$	_____	_____
1993	$1.20	$59\frac{1}{8}$	_____	_____
1994	$1.20	$50\frac{3}{8}$	_____	_____
1995	$1.35	$69\frac{1}{4}$	_____	_____

30 SECTION 1   COMMON STOCKS: AN INTRODUCTION

15. Bought 200 shares of Chrysler Corporation stock for $11\frac{1}{16}$ per share at the end of December, 1987.

Year	Annual Dividend per Share	Year-End Closing Price per Share	Dividend Yield Based Upon Closing Price	Annual Return
1988	$0.50	$12\frac{7}{8}$	_____	_____
1989	$0.58	$9\frac{1}{2}$	_____	_____
1990	$0.60	$6\frac{5}{16}$	_____	_____
1991	$0.38	$5\frac{7}{8}$	_____	_____
1992	$0.30	16	_____	_____
1993	$0.30	$26\frac{5}{8}$	_____	_____
1994	$0.45	$24\frac{1}{2}$	_____	_____
1995	$0.90	$27\frac{9}{16}$	_____	_____

COMMON STOCKS: AN INTRODUCTION   SECTION 1 **31**

16. Bought 100 shares of Delta Airlines, Incorporated stock for 48 per share on January 2, 1987.

Year	Annual Dividend per Share	Year-End Closing Price per Share	Dividend Yield Based Upon Closing Price	Annual Return
1987	$1.10	$37\frac{1}{8}$	_____	_____
1988	$1.20	$50\frac{1}{8}$	_____	_____
1989	$1.70	$68\frac{1}{4}$	_____	_____
1990	$1.20	$55\frac{3}{4}$	_____	_____
1991	$1.20	$66\frac{1}{8}$	_____	_____
1992	$1.20	$50\frac{7}{8}$	_____	_____
1993	$0.20	$54\frac{5}{8}$	_____	_____
1994	$0.20	$50\frac{1}{2}$	_____	_____
1995	$0.20	$73\frac{5}{8}$	_____	_____

Section 2
Common Stocks: Buying and Selling

As mentioned in the preceding section, common stocks are usually purchased through brokerage firms that are members of a stock exchange. There are several stock exchanges in the United States, the two largest being the New York Stock Exchange (NYSE) and the American Stock Exchange (AMEX). Stocks that are traded outside of a stock exchange are called **over-the-counter stocks (OTC)**. The largest and best-known of OTC stocks is called the National Association of Securities Dealers Automated Quotation System (NASDAQ). Most of the nation's larger newspapers list daily quotations for stocks traded through these major markets (NYSE, AMEX, NASDAQ). Two of the largest publications specializing in stock quotations are *The Wall Street Journal* (published every weekday) and *Barron's* (published once a week).

A typical newspaper listing of stock activity is shown in Table 3. The first two columns (52 weeks high and low) list the range of prices for the 52 weeks ending the previous working day (January 7 in Table 3). The sales column lists (in hundreds of shares) the number of shares that were traded that day. The daily price column lists the high and low prices for that day as well as the final price at closing. The net change lists the difference between that day's closing price and the previous day's closing price.

To save space, each corporation's name in a stock listing is given in abbreviated form. For example, the corporations listed in Table 3 (along with their ticker symbols) are: Litton (LIT), Living Centers of America (LCA), Liz Claiborne (LIZ), Lockheed Martin (LMT), Loctite (LOC), Loewen Group (LWN), Loews Corporation (LTR), Logicon (LGN), Nordic American Tanker Shipping, Limited (NAT+), Norex America Inc. (NXA), North American Vaccine (NVX), Northern Technologies (NTI), Micros-to-Mainframes (MTMC), MicroSemi Corporation (MSCC), MicroSoft (MSFT), Microtest (MTST), and Microtouch Systems (MTSI). Corporate stocks traded on the New York Stock Exchange are usually more well known than those on the American Stock Exchange and a good number of those listings under NASDAQ are technology-based companies.

TABLE 3

NYSE Wednesday January 8, 1997

52 Weeks High	52 Weeks Low	Stock	Dividend	Current Yield	P/E Ratio	Sales (100s)	Daily High	Daily Low	Daily Close	Net Change
$51\frac{1}{2}$	$40\frac{1}{2}$	Litton	...	...	14	802	$46\frac{1}{2}$	46	$46\frac{1}{4}$	$-\frac{1}{4}$
41	$20\frac{3}{4}$	LivngCtr	...	...	13	742	$28\frac{7}{8}$	$28\frac{1}{4}$	$28\frac{3}{4}$	$\frac{1}{8}$
$45\frac{1}{8}$	$26\frac{1}{4}$	LizClab	$0.45	1.1%	21	2434	$40\frac{3}{4}$	$39\frac{3}{4}$	$40\frac{1}{8}$	$\frac{1}{8}$
$96\frac{5}{8}$	73	LockhdM	$1.60	1.7%	16	3164	$92\frac{1}{2}$	$89\frac{7}{8}$	$92\frac{1}{2}$	$1\frac{3}{4}$
$61\frac{7}{8}$	$42\frac{1}{4}$	Loctite	$1.20	2.0%	24	546	$60\frac{7}{8}$	$60\frac{3}{4}$	$60\frac{3}{4}$	$-\frac{1}{8}$
$26\frac{3}{4}$	$15\frac{5}{8}$	LoewenG	$2.36	8.9%	...	74	$26\frac{5}{8}$	$26\frac{1}{2}$	$26\frac{5}{8}$	...
43	$16\frac{3}{8}$	Loewen	$0.10	0.2%	dd	977	$41\frac{1}{8}$	$40\frac{7}{8}$	$40\frac{7}{8}$	$\frac{1}{4}$
$98\frac{5}{8}$	$72\frac{1}{2}$	Loews	$1.00	1.1%	6	1401	$95\frac{3}{4}$	$94\frac{3}{8}$	95	$-1\frac{3}{4}$
$43\frac{3}{8}$	$25\frac{1}{4}$	Logicon	$0.24f	0.6%	18	185	$37\frac{3}{8}$	$36\frac{5}{8}$	37	$\frac{1}{4}$

AMEX Wednesday January 8, 1997

52 Weeks High	52 Weeks Low	Stock	Dividend	Current Yield	P/E Ratio	Sales (100s)	Daily High	Daily Low	Daily Close	Net Change
$39\frac{7}{8}$	$25\frac{3}{4}$	NY Times	$0.60f	1.6%	23	1683	$37\frac{5}{8}$	$36\frac{3}{8}$	$37\frac{5}{8}$	1
$4\frac{3}{4}$	$3\frac{1}{4}$	Nordic wt	...	...	...	50	$4\frac{1}{2}$	$4\frac{1}{2}$	$4\frac{1}{2}$	...
$35\frac{5}{8}$	$12\frac{1}{4}$	▲Norex	...	...	dd	8	$36\frac{1}{2}$	$35\frac{7}{8}$	$36\frac{1}{2}$	$\frac{7}{8}$
$28\frac{3}{8}$	$11\frac{3}{4}$	NA VACC	...	...	dd	105	$24\frac{3}{4}$	$24\frac{1}{2}$	$24\frac{5}{8}$	$-\frac{3}{8}$
7	$4\frac{3}{4}$	Nthn Tch	$0.12f	2.0%	12	37	$6\frac{1}{4}$	$5\frac{7}{8}$	6	$\frac{1}{4}$

NASDAQ Wednesday January 8, 1997

52 Weeks High	52 Weeks Low	Stock	Dividend	Current Yield	P/E Ratio	Sales (100s)	Daily High	Daily Low	Daily Close	Net Change
$6\frac{3}{4}$	$2\frac{1}{8}$	Micros To	...	...	dd	226	$3\frac{1}{16}$	$2\frac{15}{16}$	$3\frac{1}{16}$	$\frac{1}{8}$
15	$7\frac{3}{8}$	▲MicSem	...	...	15	3674	$15\frac{3}{8}$	$14\frac{7}{8}$	$15\frac{1}{8}$	$\frac{1}{8}$
$86\frac{1}{8}$	40	Microsft s	...	...	47	46101	$85\frac{1}{8}$	$83\frac{3}{8}$	85	$\frac{5}{8}$
$81\frac{1}{2}$	$79\frac{3}{4}$	Micsft pf	...	...	...	738	$80\frac{1}{4}$	80	$80\frac{1}{8}$	$\frac{1}{8}$
$12\frac{1}{4}$	5	Micrtest	...	...	20	306	$9\frac{7}{8}$	$9\frac{5}{8}$	$9\frac{3}{4}$	...
$29\frac{3}{4}$	$11\frac{1}{4}$	MictchSy	...	...	38	77	$24\frac{1}{4}$	24	$24\frac{1}{2}$	$\frac{1}{4}$

Some of the entries in newspaper stock listings are prefaced (or followed) by code letters or symbols. Different newspapers use different codes and normally a newspaper will print its codes along with the listing. The codes for the letters and symbols used in Table 3 are as follows.

dd	Loss in last 12 months
f	Current annual rate, which was decreased by most recent dividend announcement
▲	Yesterday's high was greater than 52-week high
pf	Preferred stock issue
s	Stock has split by at least 20 percent within the last year, and prices have been adjusted accordingly.

EXAMPLE 1 Reading daily stock quotations

a. What was the lowest price per share of Lockheed Martin common stock during the 52 weeks prior to January 8, 1997?

b. How many shares of Loews Corporation common stocks were traded on January 8, 1997?

c. What was the closing price of New York Times common stock on January 8, 1997?

d. What was the closing price of New York Times common stock on January 7, 1997?

SOLUTION

a. The lowest price per share of Lockheed Martin common stock during the 52 weeks prior to January 8, 1997 was 73 or $73.00 per share.

b. The number of shares of Loews Corporation common stock traded on January 8, 1997 was 1401 (in hundreds of shares). That is,

Number of Shares = (1401)(100) = 140,100.

c. The closing price of New York Times common stock on January 8, 1997 was $37\frac{5}{8}$ or $37.625 per share.

d. Since the closing price for New York Times common stock was up 1 on January 8, 1997, the closing price on January 7 was 1 less than $37\frac{5}{8}$ or

Closing Price on January 7 = $37\frac{5}{8} - 1 = 36\frac{5}{8}$. ◆

Try one!

a. What was the lowest price per share of Loctite common stock during the 52 weeks prior to January 8, 1997?

Answer: _____

b. How many shares of North American Vaccine common stocks were traded on January 8, 1997?

Answer: _____

c. What was the closing price of Microtest common stock on January 8, 1997?

Answer: _____

d. What was the closing price of Microtest common stock on January 7, 1997?

Answer: _____

Stocks are usually bought and sold in **round lots** of 100 shares, 200 shares, 300 shares, etc. Lot sizes of less than 100 shares are called **odd lots**. Each time an investor buys or sells stocks, he or she must pay a **broker's commission**. This commission depends upon the type of broker (full service or discount), the price per share, and the number of shares. In addition to the regular commission, investors may pay an extra fee when dealing in odd lots.

Commissions vary greatly from one brokerage firm to another. However, all broker's commissions share the feature that the percent of the commission goes down as the purchase price goes up. A typical full service broker's commission will usually be a minimum of about $50.00. Therefore, for a transaction of $500.00, the broker's commission percentage would be about 10%. Broker's commissions cover the cost to the broker for handling the trade as well as brokerage overhead costs.

Up until 1975, the minimum rates for all brokers were set by the Securities and Exchange Commission (SEC). Since that time, the Securities and Exchange Commission has allowed brokerage firms to compete on a commission basis. As a result of this competition, brokerage firms have divided into two categories: **discount brokerage firms**, and **full-service brokerage firms**. (Some of these brokerage firms even operate online services where clients can trade stocks over the internet!) Discount brokerage firms buy and sell stocks for their clients but provide little or no additional service. Such firms can save money (in commissions) for investors who know precisely which stocks they wish to buy or sell. Investors who want help in keeping up with their stocks would do better to pay the higher commission rates of a full-service broker. A good full-service broker follows the stocks of each client and makes suggestions for buying or selling.

In Section 1, we defined the *total income* and *annual return* for stocks to be

Total Income $= SP + D - PP$

$$\text{Annual Return} = \frac{\text{Ending Price} - \text{Beginning Price} + \text{Dividends}}{\text{Beginning Price}}$$

where *SP* stands for selling price, *D* stands for dividends, and *PP* stands for purchase price. Remember that the ending price *EP* is the price of the stock at the end of the year (also corresponds to the selling price) and the beginning price *BP* is the price of the stock at the beginning of the year (also corresponds to the purchase price). For the **net annual return** on the purchase and sale of a stock, we include the broker's commissions *BC*, both at the time of purchase and at the time of sale.

$$\text{Net Annual Return} = \frac{EP - BP + \text{Dividends} - BC}{BP}$$

EXAMPLE 2   Finding the net annual return on common stock

Mike Henderson purchased 100 shares of Loews Corporation common stocks in January of 1996 at $79\frac{1}{2}$. If Mike sold (a year later) at the closing price on January 8, 1997, what was his net annual return? (Assume that Mike paid a 1.0% commission for buying and a 0.4% commission for selling.)

SOLUTION

The purchase price for Mike's 100 shares at $79\frac{1}{2}$ per share was

Purchase Price = (100)($79.50) = $7950.00

and the 1.0% commission on this purchase was

Purchase Commission = (0.01)($7950.00) = $79.50.

From Table 3, we see that Mike earned dividends of $1.00 per share during the year. Thus, his total dividends were

Dividends = (100)($1.00) = $100.00.

Since the closing price on January 8, 1997 was 95 per share, Mike's selling price was

Selling Price = (100)($95.00) = $9500.00

and the commission on this sale was

Selling Commission = (0.004)($9500.00) = $38.00.

COMMON STOCKS: BUYING AND SELLING

Recalling the formula for calculating the net annual return,

$$\text{Net Annual Return} = \frac{EP - BP + \text{Dividends} - BC}{BP}$$

we will first determine the numerator (top number) of the fraction. This number represents the *net income* on Mike's original investment.

	$9500.00	ending (selling) price
−	7950.00	beginning (purchase) price
+	100.00	dividends
−	79.50	purchase commission
−	38.00	selling commission
	$1532.50	net income

By dividing the net income for one year by the purchase price, we determine the net annual return.

$$\text{Net Annual Return} = \frac{\$1532.50}{\$7950.00} \approx 0.193 = 19.3\%$$ ◆

Try one!

Bradley Peterson purchased 200 shares of Logicon common stock in January of 1996 at $29\frac{1}{4}$. If Bradley sold (a year later) at the closing price on January 8, 1997, what was his net annual return? (Assume that Bradley paid a 0.8% commission for buying and a 0.5% commission for selling.)

Answer: _____

Note that in Example 2, Mike's combined broker's commissions of $117.50 were fairly insignificant when compared with his net income. However, remember that commission rates are higher for small investments and in some cases may turn a reasonable return into a poor one as illustrated in the next example.

EXAMPLE 3 Finding the net return on common stock

Muriel Berkley bought 100 shares of Northern Technologies common stock at $5\frac{1}{8}$ in January of 1996. If she sold at the closing price on January 8, 1997, what was her net annual return? (Assume that Muriel paid a $45.00 commission for both buying and selling.)

SOLUTION

From Table 3, we see that Muriel sold at 6 and received dividends of $0.12 per share during the year. Thus, her net income for the year was

	$600.00	ending (selling) price:	(100)($6.00)
−	512.50	beginning (purchase) price:	(100)($5.125)
+	12.00	dividends:	(100)($0.12)
−	45.00	purchase commission	
−	45.00	selling commission	
	$9.50	net income	

and her net annual return was

$$\text{Net Annual Return} = \frac{\$9.50}{\$512.50} \approx 0.019 = 1.9\%.$$

Note that Muriel's annual return (before commissions were paid) was

$$\text{Annual Return} = \frac{EP - BP + \text{Dividends}}{BP} = \frac{\$99.50}{\$512.50}$$

$$\approx 0.194 = 19.4\%$$

which is a very good annual return on an investment. However, after paying broker's commissions totaling $90.00, Muriel's net income was only $9.50. She could have done significantly better by putting her money in a bank savings account. ◆

COMMON STOCKS: BUYING AND SELLING   SECTION 2   **41**

Try one!

Sharon and Robert Byerston bought 100 shares of Loctite Corporation common stock at $47\frac{1}{2}$ in January of 1996. If they sold at the closing price on January 8, 1997, what was their net annual return? (Assume that the Byerstons paid a $50.00 commission for buying and a $65.00 commission for selling.)

Answer: _____

In most corporations, the board of directors likes to keep the price per share of common stock within a range affordable to the average investor. To do this a corporation may **split** its common stock. Splits usually occur on a basis of two-for-one, but may also occur in other multiples such as three-for-one or three-for-two. When a split occurs, stockholders are issued new shares in proportion to the number of shares they currently own. For example, if a stockholder owned 200 shares and a two-for-one split was ordered, that stockholder would receive 200 additional shares for a total of 400 shares. Of course, the price per share would drop in proportion to the split, leaving the stockholder's total worth in the split stocks unaffected. For example, a stock selling for $50.00 per share would usually sell for around $25.00 per share immediately following a two-for-one split.

EXAMPLE 4   Finding the selling price after a stock split

Connie and Bill Meyers purchased 300 shares of Logicon common stock in early 1995 for $37\frac{1}{4}$ per share. In September 1995, Logicon common stock split two-for-one. Connie and Bob then sold their shares at the closing price on January 8, 1997.

a. What was the Meyers' purchase price for this stock?

b. What was the Meyers' selling price for this stock?

SOLUTION

From Table 3, we see that the closing price on January 8, 1997 was 37 per share. At first it might seem that Connie and Bill lost money since they bought at $37\frac{1}{4}$ and sold at 37. However, they only paid for 300 shares and after the two-for-one split, they owned $(300)(\frac{2}{1}) = 600$ shares.

a. The Meyers' purchase price was

   Purchase Price = (300)($37.25) = $11,175.00.

b. The Meyers' selling price was

   Selling Price = (600)($37.00) = $22,200.00.   ◆

Try one!

Hal Dobbs purchased 200 shares of Loews Corporation common stock in 1992 for $103\frac{1}{2}$ per share. In October 1995, Logicon Corporation common stock split two-for-one. Hal then sold his shares at the high price on January 8, 1997.

a. What was Bill's purchase price for this stock?

Answer: _____

b. What was Bill's selling price for this stock?

Answer: _____

COMMON STOCKS: BUYING AND SELLING   SECTION 2   43

A second way that a stockholder's number of shares can increase is by a **stock dividend**. Sometimes this is done in place of a cash dividend and sometimes it accompanies a cash dividend. Stock dividends are always given on a percentage basis. A two percent stock dividend means that each stockholder receives two additional shares for each 100 owned. (Although there are technical differences, a 100% stock dividend is comparable to a two-for-one split.)

EXAMPLE 5 Finding the selling price after a stock dividend

Warren Richards bought 800 shares of Vishay Intertechnology, Incorporated common stock in late 1993 at $29\frac{3}{4}$ per share. In May, 1994, Warren received a 5% stock dividend; in March 1995, a 5% stock dividend; and in May 1995, a 50% stock dividend. Finally, in early 1996 Warren sold at 30. (There were no *cash* dividends paid during this time period.)

a. What was Warren's purchase price for this stock?

b. What was Warren's selling price for this stock?

SOLUTION

a. Warren's purchase price was

Purchase Price = (800)($29.75) = $23,800.00.

b. After the first 5% stock dividend, Warren's shares were increased by 5%, leaving him with

800 + (0.05)(800) = 840 shares.

After the second 5% stock dividend, Warren's shares were increased by an additional 5%, leaving him with

840 + (0.05)(840) = 882 shares.

After the 50% stock dividend, Warren had

882 + (0.50)(882) = 1323 shares.

Therefore, Warren's selling price was

Selling Price = (1323)($30.00) = $39,690.00. ◆

Try one!

The Williams family bought 5000 shares of Bell Industries, Incorporated common stock in early 1993 at $16\frac{1}{2}$ per share. In July, 1993, the Williams' received a 4% stock dividend; in October, 1994, a 5% stock dividend; and in May 1995, a 5% stock dividend. Finally, in early 1996 they sold at 18.

a. What was the Williams' purchase price for this stock?

Answer: _____

b. What was the Williams' selling price for this stock?

Answer: _____

Success stories like Warren's in Example 5 have an enticing effect upon investors. In roughly two and one half years, Warren's investment of $23,800.00 increased by more than 60 percent. In order to do this well on an investment, some investors are willing to borrow money to purchase stocks. If this money is borrowed from a brokerage firm, the investor is said to be **buying on margin**. The Federal Reserve Board governs the percentage an investor can borrow when buying on margin and this **margin rate** varies from time to time. In past years, investors have been allowed to borrow 50 percent of the purchase price of stocks. Investors can also borrow from banks and other financial institutions to buy stocks.

An investor who chooses to buy on margin by borrowing from a brokerage or other financial institution will be required to pay **interest** on the borrowed funds. The amount of interest paid depends upon the amount borrowed (the **principal**), the percentage **rate** charged by the brokerage or financial institution, and the length of **time** (in years) that the investor has the loan. The formula for determining the amount of interest owed is as follows.

$$\text{Interest} = PrT$$

where

P = principal, r = rate, and T = time (in years).

The interest paid when buying stock on margin has an effect on the net income realized upon sale of the stock. We define the **net income (on margin)** as

$$\text{Net Income (on Margin)} = EP - BP + D - C - \text{Interest}$$

where D = dividends, and C = broker's commissions.

EXAMPLE 6   Buying on margin

Suppose that Warren Richards (see Example 5) used his $23,800.00 to purchase (on margin) 1600 shares of Vishay Intertechnology, Incorporated common stock in late 1993 at $29\frac{3}{4}$ per share. As in Example 5, Warren sold in early 1996, two and one half years later at 30. Remember, Warren received no cash dividends on this stock.

a. What was the purchase price of this stock?

b. How much of this purchase price did Warren borrow?

c. At a six percent annual percentage rate for two and one half years, how much interest did Warren pay his brokerage firm?

d. What was Warren's selling price for this stock?

e. What was Warren's net income on margin? Assume that Warren paid $238.00 commission on the purchase and $317.00 commission on the sale of this stock.

SOLUTION

a. The purchase (beginning) price was

Purchase Price = (1600)($29.75) = $47,600.00.

b. Since Warren invested only $23,800.00, he borrowed

Margin Loan = $47,600.00 − $23,800.00 = $23,800.00.

c. At six percent interest for two and one half years, Warren paid a total interest of

$I = PrT$ = ($23,800.00)(0.06)(2.5) = $3570.00.

d. After the stock dividends of 5%, 5%, and 50%, Warren's 1600 shares increased as follows.

1600 + (0.05)(1600) = 1680 shares
1680 + (0.05)(1680) = 1764 shares
1764 + (0.50)(1764) = 2646 shares

Therefore, Warren's selling (ending) price was

Selling Price = (2646)($30.00) = $79,380.00.

e. Warren's net income on margin was

Net Income (on Margin) = $EP - BP + D - C -$ Interest

$\quad$ = $79,380.00 − $47,600.00 + $0.00
$\quad\quad$ − $555.00 − $3570.00

$\quad$ = $27,655.00 ◆

Keep in mind that when buying on margin, the net income is reduced by the amount of interest paid. Furthermore, when you buy on margin, not only do you have to pay interest, but since you are buying and selling more shares, your commission will also be higher. Even so, investors who buy on margin hope that the increased selling price will compensate for increased expenses.

COMMON STOCKS: BUYING AND SELLING

Try one!

The Williams family bought (on margin) 10,000 shares of Bell Industries, Incorporated common stock in early 1993 at $16\frac{1}{2}$ per share. They had enough cash to buy only 5000 shares and borrowed the rest. In July, 1993 the Williams received a 4% stock dividend; in October, 1994, a 5% stock dividend; and in May 1995, a 5% stock dividend. Finally, in early 1996 (three years later) they sold at 18. They paid a 0.5% commission for both buying and selling. Assume that the Williams family received a total of $1000.00 in cash dividends during this three-year period.

a. What was the Williams' purchase price for this stock?

Answer: _____

b. How much of this purchase price did they borrow?

Answer: _____

c. At a 5% annual percentage rate for three years, how much interest did they pay the brokerage firm?

Answer: _____

d. What was the Williams' selling price for this stock?

Answer: _____

e. What was the net income (on margin) for this investment?

Answer: _____

Buying on margin involves more risk than buying with cash. The problem is that while buying on margin can magnify a gain on a fixed investment, it can just as easily magnify a loss. If the price of a stock purchased on margin falls too far below the original purchase price, the brokerage firm will issue a **margin call** asking the investor to put up more cash in order to protect the firm's loan. The last example in this section shows this situation as it happened to an investor in the 1970s.

EXAMPLE 7 Finding the net income when buying on margin

Between mid-1972 and mid-1973, Coca-Cola Bottling Company of New York common stock fell from a high of over 33 to a low of less than 17. Sharon Reed decided that this was a good time to buy and purchased (on margin) 500 shares at $18\frac{1}{4}$ in mid-1973. (Sharon invested $6000.00 and borrowed $3125.00 of the purchase price.) Things were going well for Sharon through the early fall of 1973 and then the price per share began to decline as follows.

Date (Fridays)	November				December				
	2	9	16	23	1	7	14	21	28
Price	$24\frac{3}{4}$	$24\frac{1}{4}$	$23\frac{1}{4}$	$21\frac{1}{4}$	19	$18\frac{3}{4}$	$16\frac{1}{4}$	$13\frac{3}{4}$	$9\frac{1}{8}$

Near the end of December, Sharon's brokerage firm issued a margin call on Sharon's stock, asking Sharon to put up additional cash. After Sharon failed to do so, the firm sold her stock at $9\frac{1}{8}$ per share. Find Sharon's net income using the following information.

Dividends (two quarters)	$0.18 per share
Interest (six months)	9% annual percentage rate
Purchase Commission	0.9% of purchase price
Selling Commission	1.2% of selling price

COMMON STOCKS: BUYING AND SELLING

SOLUTION

Sharon's dividends for two quarters amounted to

$$\text{Dividends} = (500)(\$0.18) = \$90.00.$$

The interest on the $3125.00 loan was

$$\text{Interest} = PrT = (\$3125.00)(0.09)(0.5) \approx \$140.63.$$

The purchase commission was

$$\text{Purchase Commission} = (\$9125.00)(0.009) \approx \$82.13.$$

The selling price was

$$\text{Selling Price} = (500)(\$9.125) = \$4562.50.$$

The selling commission was

$$\text{Selling Commission} = (\$4562.50)(0.012) = \$54.75.$$

Therefore, Sharon's net income was

	$4562.50	selling (ending) price
−	9125.00	purchase (beginning) price
+	90.00	dividends
−	82.13	purchase commission
−	54.75	selling commission
−	140.63	interest
	− $4750.01	net income (In this case it was a net loss.)

(In other words, Sharon lost $4750.01 of her $6000.00 investment.) ◆

Important Terms

broker's commission	net income (on margin)
buying on margin	odd lot
closing price	over-the-counter stocks
discount brokerage firm	P, principal
full-service brokerage firm	r, rate
interest	round lot
margin call	stock dividend
margin rate	stock split
net annual return	T, time in years

Important Formulas

$$\text{Annual Return} = \frac{\text{Ending Price} - \text{Beginning Price} + \text{Dividends}}{\text{Beginning Price}}$$

Net Income = Selling Price − Purchase Price + Dividends

$$\text{Net Annual Return} = \frac{EP - BP + D - BC}{BP}$$

Interest = PrT

Net Income $\binom{\text{On}}{\text{Margin}}$ = $SP + D - PP -$ Commissions $-$ Interest

CONSUMER HINTS

- The larger and more well established a corporation is, the less your risk. "Hot tips" about newly emerging corporations usually reach the average investor in exaggerated form. As illustrated in this and the preceding section, it is not necessary to buy obscure stocks in order to reap good returns.

COMMON STOCKS: BUYING AND SELLING SECTION 2 **51**

SECTION 2   EXERCISES

In Exercises 1–8, use the stock quotations found in Table 3 on page 34.

1. How many shares of Litton stock were traded on January 8, 1997?

   Answer: _____

2. How many shares of Lockheed Martin stock were traded on January 8, 1997?

   Answer: _____

3. Find the cost of the following stock purchases (ignore the broker's fees).

   a. 500 shares of Loctite stock at the closing price on January 8, 1997.

   Answer: _____

   b. 100 shares of Loews Corporation stock at the high price on January 8, 1997.

   Answer: _____

   c. 100 shares of Loews Corporation stock at the low price on January 8, 1997.

   Answer: _____

4. Find the cost of the following stock purchases (ignore broker's fees).

a. 200 shares of New York Times common stock at the closing price on January 8, 1997.

Answer: _____

b. 200 shares of New York Times common stock at the high price on January 8, 1997.

Answer: _____

c. 500 shares of Northern Technologies stock at the low price on January 8, 1997.

Answer: _____

5. What was the closing price of Liz Claiborne stock on January 7, 1997?

Answer: _____

6. What was the closing price of North American Vaccine stock on January 7, 1997?

Answer: _____

COMMON STOCKS: BUYING AND SELLING　　　　　　　　　　　　　　　　　SECTION 2　**53**

7. What was the difference between the high and low prices of Logicon common stock for the 52 weeks prior to January 8, 1997?

Answer: _____

8. What was the difference between the high and low prices of Microsoft Preferred stock for the 52 weeks prior to January 8, 1997?

Answer: _____

In Exercises 9–12, determine the broker's commission.

9. Buy 100 shares of Litton stock at 46 per share with a broker's commission of 1.2%.

Answer: _____

10. Sell 500 shares of Nordic American Tanker Shipping stock at $4\frac{1}{2}$ with a broker's commission of 2.3%.

Answer: _____

11. Sell 200 shares of New York Times stock at 37 and buy 400 shares of Living Centers of America stock at $28\frac{3}{8}$ when the broker's commission on each transaction is 0.7%.

Answer: _____

12. Sell 300 shares of Northern Technologies common stock at $6\frac{1}{8}$ and buy 400 shares of Loctite stock at $60\frac{3}{4}$ when the commission on each transaction is 0.9%.

Answer: _____

13. George Duncan bought 100 shares of Delta Airlines stock for $45\frac{1}{2}$ per share in 1986 and sold it for $73\frac{5}{8}$ per share at the end of 1995. The average dividend per share over the ten years was $0.92. Assuming that the broker's commission rate for buying was 1.2% and for selling was 0.6%, find George's net income.

Answer: _____

COMMON STOCKS: BUYING AND SELLING SECTION 2 **55**

14. Gloria Dever bought 100 shares of Aetna, Incorporated stock for 53 per share in 1986 and sold it for $67\frac{7}{8}$ per share at the end of 1995. The average dividend per share over the ten years was $2.75. Assuming that the broker's commission for buying and selling was 1.0%, find Gloria's net income.

Net Income: _____

15. Carla Morgan bought 200 shares of Eastern Utilities Associates stock for $21\frac{1}{8}$ per share in 1988 and sold it for $23\frac{5}{8}$ per share at the end of 1995. The average dividend per share over the eight year period was $1.85. Find Carla's net income if the broker's commission rate for buying was $50.00 and for selling was $59.60.

Net Income: _____

16. Andrea Smith bought 200 shares of a cosmetic company stock at $27\frac{1}{2}$. Two years later, the stock split two for one. Andrea sold the stock at $24\frac{1}{4}$ three years after the split after receiving a total dividend of $445.00. Find Andrea's total income on this stock.

Answer: _____

17. Gary Hartman bought 1000 shares of an industrial corporation's stock at $9\frac{3}{4}$ and about four years later the corporation's stock split two-for-one. A year later Gary sold the stock at 10 after receiving a total of $575.00 in dividends. Find Gary's total income on this stock.

Answer: _____

COMMON STOCKS: BUYING AND SELLING   SECTION 2 **57**

In Exercises 18–20, find the purchase price and the selling price for the given stock transaction.

18. Bought 100 shares of Intel Corporation stock at $26\frac{1}{8}$ in 1977. There was a five-for-four split in September 1978 and the stock was sold at $30\frac{7}{8}$ in 1979.

Purchase Price:_____

Selling Price: _____

19. Bought 500 shares of Nike, Incorporated stock at $29\frac{1}{2}$ in 1988 and sold it at 69 in January of 1996. In 1990 the stock split two-for-one and late in 1995 it split two-for-one once again.

Purchase Price:_____

Selling Price: _____

20. Bought 100 shares of McDonald's Corporation stock in 1986 for $109\frac{3}{4}$ and sold in late 1995 for 46. There was a three-for-two split in June of 1987 and there were two-for-one splits in June of 1989 and May of 1994.

Purchase Price:_____

Selling Price: _____

21. Rick Ferguson purchased (on margin) 1000 shares of a particular life and casualty company stock at 11 in 1990 (Rick invested $7000.00 and borrowed the remaining $4000.00 of the purchase price). In 1994, Rick sold at $28\frac{1}{2}$ after receiving a total of $4370.00 in dividends. Find the total income if Rick has to pay 10% annual percentage rate for $3\frac{1}{2}$ years on the amount borrowed. Assume that the broker's commission was 1% for buying and again for selling.

Answer: _____

Section 3
Corporate Bonds

Corporations, like people, occasionally need to borrow money and one way to do this is to issue **corporate bonds**. Simply stated, a bond is an IOU in which the issuer of the bond agrees to repay the loan after a certain period of time. Most corporate bonds are likely to be **debenture bonds**; that is, they are not backed by collateral. **Nondebenture bonds** are backed by some form of corporate property. When new bonds are issued, the investment banker usually places a notice in several widely distributed magazines or newspapers. Figure 8 shows such an advertisement (called a tombstone) for Tenneco bonds.

This announcement is neither an offer to sell nor a solicitation of an offer to buy any of these Securities.

The offer is made only by the Prospectus.

$250,000,000

Tenneco Inc.

$9\frac{1}{2}$% Debentures due 2004

*Interest payable
June 15 and December 15*

Price $99\frac{3}{8}$% and Accrued Interest

June 11, 1979

Figure 8

When purchasing a corporate bond, five items are of primary importance.

1. The *financial condition* of the **bond issuer** determines the risk involved. If a bond is purchased from a corporation that subsequently declares bankruptcy, the owner of the bond may lose part

59

(or all) of his or her investment. Government bonds or bonds issued by major corporations offer the least risk. Bonds issued by new corporations or by corporations with weak financial outlooks involve greater risk. (In Figure 8, Tenneco Incorporated is the bond issuer.)

2. The **par value** of a bond is the amount the issuer agrees to pay the bondholder at the time the bond reaches maturity. The par value of most bonds is $1000.00.

3. The **maturity date** of the bond is the date the issuer agrees to pay back the loan. The number of years between the date of issue and the maturity date is called the *term* of the bond. Bond terms vary from five or ten up to 50 years. (In Figure 8, the maturity date is 2004, and the term is 25 years.)

4. The **coupon rate** of the bond is the annual interest rate (based on the par value) that the bond issuer agrees to pay the bondholder. Most bonds pay interest twice per year but some pay only once per year. (In Figure 8, the annual percentage rate is $9\frac{1}{2}\%$ with interest paid semiannually.)

5. The price may be different from the par value. When a bond (new or old) is purchased, the price is listed as a percentage of the par value. For example, the bonds described in Figure 8 have a price of $993.75 ($99\frac{3}{8}\%$ of the par value of $1000.00).

EXAMPLE 1   Finding the interest for a bond

Chad and Renee Brewer purchased five Tenneco Inc. bonds with a par value of $1000.00 each and interest dates of June 15 and December 15. If the coupon rate is $9\frac{1}{2}\%$, how much interest will Chad and Renee receive from these bonds every six months?

SOLUTION

For each bond, the interest for six months (one-half of a year) is

$$\text{Interest} = PrT = (\$1000.00)(0.095)(\tfrac{1}{2}) = \$47.50.$$

Since Chad and Renee own five bonds, their six-month interest checks will amount to

$$\text{Total Interest} = (5)(\$47.50) = \$237.50.$$ ◆

CORPORATE BONDS

Try one!

Steven and Ruby Thompson purchased seven Tenneco Inc. bonds with a par value of $1000.00 each and interest dates of June 15 and December 15. If the coupon rate is $9\frac{1}{2}\%$, how much interest will Steven and Ruby receive from these bonds every six months?

Answer: _____

When a newly issued bond is held until maturity, the transaction is similar to buying a savings certificate at a bank. Thus, bonds held to maturity represent a fairly conservative, long-term investment. In practice, however, many people do not keep bonds until maturity. If an investor wishes to sell a bond before its maturity date, the bond can be placed on the open market (through the assistance of a stockbroker) and sold to another investor. The catch is that if the bond's coupon rate is low, or if the issuer of the bond appears to be in financial difficulty, there may not be anyone willing to pay the par value of the bond. The price an investor is willing to pay for a bond (new or old) is called its **current price**. The January 1997 current prices of several corporate bonds are listed in Table 4.

TABLE 4

Current Prices of Selected Corporate Bonds (January 1997)

Issuer	Rate	Par Value	Maturity Date	Date of Issue	Current Price*	
Cambridge Electric Light	$7\frac{3}{4}\%$	$1000.00	2002	1972	$1002.50	($100\frac{1}{4}$)
Data General	$8\frac{3}{8}\%$	$1000.00	2002	1977	$995.00	($99\frac{1}{2}$)
Exxon Corporation	$8\frac{1}{4}\%$	$1000.00	1999	1989	$1050.00	(105)
Florida Power and Light	9.8%	$1000.00	2018	1988	$1085.00	($108\frac{1}{2}$)
Ford Motor Company	9.15%	$1000.00	2004	1979	$1012.50	($101\frac{1}{4}$)
General Electric	$7\frac{7}{8}$	$1000.00	1998	1991	$1028.75	($102\frac{7}{8}$)
J.C. Penney	$7\frac{3}{8}\%$	$1000.00	2004	1994	$1023.75	($102\frac{3}{8}$)
Owens-Corning	5.79%	$1000.00	2005	1985	$980.00	(98)
Proctor and Gamble	8.0%	$1000.00	2029	1989	$1100.00	(110)
Tenneco, Inc.	$7\frac{1}{4}\%$	$1000.00	2025	1995	$938.75	($93\frac{7}{8}$)

*Note that a current price of $100\frac{1}{4}$ means $100\frac{1}{4}\%$ of par value.

As with stocks, we define the **total income** on bonds to be

Total Income = Selling Price + Interest − Purchase Price.

EXAMPLE 2   Finding the total income on bonds

Roger Cooper invested $5000.00 in five Owens-Corning bonds in 1985 (see Table 4). If the bonds paid a 5.79% coupon rate and Roger sold the bonds in 1997 for $980.00 each, what was his total income? (Assume that Roger collected 12 full years of interest on the bonds.)

SOLUTION

Roger's interest per bond over the 12 years amounted to

Interest = PrT = ($1000.00)(0.0579)(12) = $694.80.

Thus, for the five bonds, his total interest was

Total Interest = (5)($694.80) = $3474.00.

Since Roger sold the bonds for $980.00 each, his selling price was

Selling Price = (5)($980.00) = $4900.00.

Therefore, his total income was

$4900.00	selling price
− 5000.00	purchase price
+ 3474.00	interest
$3374.00	total income.

◆

EXAMPLE 3 Finding the total income on bonds

Christine Miller's parents invested $5000.00 in five Ford Motor Company bonds for Christine in 1979 (see Table 4). If the bonds paid a 9.15% coupon rate and Christine sold the bonds to pay for college tuition in 1997 for $1012.50 each, what was her total income? (Assume that Christine collected 18 full years of interest on the bonds.)

SOLUTION

Christine's interest per bond over the 18 years amounted to

Interest = PrT = ($1000.00)(0.0915)(18) = $1647.00.

Thus, for the five bonds, her total interest was

Total Interest = (5)($1647.00) = $8235.00.

Since Christine sold the bonds for $1012.50 each, her selling price was

Selling Price = (5)($1012.50) = $5062.50.

Therefore, Christines's total income was

$5062.50	selling price
− 5000.00	purchase price
+ 8235.00	interest
$8297.50	total income.

◆

Try one!

Albert Garfield invested $7000.00 in seven Cambridge Electric Light bonds in 1972 (see Table 4). If the bonds paid a $7\frac{3}{4}\%$ coupon rate and Albert sold the bonds in 1997 for $1002.50 each, what was his total income? (Assume that Albert collected 25 full years of interest on the bonds.)

Answer: _____

Figure 9 illustrates the correlation between inflation and corporate bond rates of return from 1965 to 1994. (Note that in Example 2, the bond was purchased during a period of lower inflation, and in Example 3 the bond was purchased at a time of higher inflation.)

CORPORATE BONDS

Figure 9
Corporate Bonds and Inflation (1965–1994)

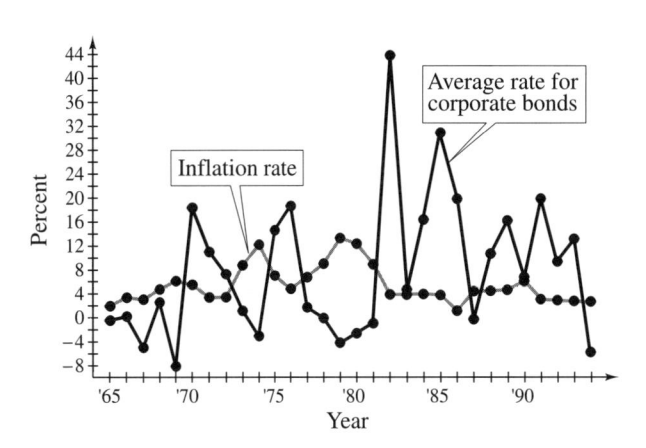

What determines the current price of bonds? One factor is the **current yield**, which is the ratio of the annual interest to the current price.

r = annual coupon rate

P = par value

p = current price

$$\text{Current Yield} = R = \frac{\text{Annual Interest}}{\text{Current Price}} = \frac{Pr}{p}$$

Calculator Hints

$R = \dfrac{Pr}{p}$

To find the current yield using your calculator, apply the following steps.

1. Multiply P by r.
2. Divide the answer to step one by p.

EXAMPLE 4   Finding the current yield of a bond

Find the current yield in January 1997 for the first three bonds listed in Table 4.

SOLUTION

a. For the Cambridge Electric Light bond, we have $r = 0.0775$, $P = \$1000.00$, and $p = \$1002.50$. Thus, the current yield is

$$R = \frac{Pr}{p} = \frac{(\$1000.00)(0.0775)}{\$1002.50} \approx 0.077 = 7.7\%.$$

EXAMPLE 4

Steps	Display
1000 × 0.0775 =	77.5
÷ 1002.50 =	0.077306733

b. For the Data General bond, we have $r = 0.08375$, $P = \$1000.00$, and $p = \$995.00$. Thus, the current yield is

$$R = \frac{Pr}{p} = \frac{(\$1000.00)(0.08375)}{\$995.0} \approx 0.084 = 8.4\%.$$

c. For the Exxon Corporation bond, we have $r = 0.0825$, $P = \$1000.00$, and $p = \$1050.00$. Thus, the current yield is

$$R = \frac{Pr}{p} = \frac{(\$1000.00)(0.0825)}{\$1050.00} \approx 0.079 = 7.9\%. \quad \blacklozenge$$

Try one!

Find the current yield in January 1997 for the next three bonds listed in Table 4.

a. Florida Power and Light

Answer: _____

b. Ford Motor Company

Answer: _____

c. General Electric

Answer: _____

CORPORATE BONDS

Table 5 gives the current yield for each of the bonds listed in Table 4.

TABLE 5

Current Yield for Selected Corporate Bonds (January 1997)

Issuer	Rate	Par Value	Maturity Date	Current Price	Current Yield
Cambridge Electric Light	$7\frac{3}{4}\%$	$1000.00	2002	$100\frac{1}{4}$	7.7%
Data General	$8\frac{3}{8}\%$	$1000.00	2002	$99\frac{1}{2}$	8.4%
Exxon Corporation	$8\frac{1}{4}\%$	$1000.00	1999	105	7.9%
Florida Power and Light	9.8%	$1000.00	2018	$108\frac{1}{2}$	9.0%
Ford Motor Company	9.15%	$1000.00	2004	$101\frac{1}{4}$	9.0%
General Electric	$7\frac{7}{8}$	$1000.00	1998	$102\frac{7}{8}$	7.7%
J.C. Penney	$7\frac{3}{8}\%$	$1000.00	2004	$102\frac{3}{8}$	7.2%
Owens-Corning	5.79%	$1000.00	2005	98	5.9%
Proctor and Gamble	8.0%	$1000.00	2029	110	7.3%
Tenneco, Inc.	$7\frac{1}{4}\%$	$1000.00	2025	$93\frac{7}{8}$	7.7%

From Example 4 and Table 5 we can see that the current yield tells just a small part of the story behind the current price of a bond since it only represents the income with respect to the price paid for a bond. Another measure which takes into consideration the potential for future gains and losses over the lifetime of the bond is called the **yield to maturity**.

We define the yield to maturity to be the average annual rate of return on purchase price. A formula for approximating this yield is below.

N = number of years to maturity

r = annual coupon rate

P = par value

p = current price

$$\text{Yield to Maturity} = R^* \approx \frac{NPr + P - p}{N\left(\frac{P+p}{2}\right)}$$

Calculator Hints

$$R^* \approx \frac{NPr + P - p}{N\left(\frac{P+p}{2}\right)}$$

To find the yield to maturity for a bond using your calculator, apply the following steps:

1. Add P to p and divide by 2.
2. Multiply by N and store.
3. Multiply N by P and r.
4. Add P and subtract p.
5. Divide by memory.
6. Round to the nearest tenth of a percent.

EXAMPLE 5   Approximating the yield to maturity

Approximate the yield to maturity for the following bonds purchased in January 1997.

a. J.C. Penney: par value ($1000.00), annual percentage rate ($7\frac{3}{8}\%$), maturity date (2004), price ($1023.75)

b. Tenneco, Incorporated: par value ($1000.00), annual percentage rate ($7\frac{1}{4}\%$), maturity date (2025), price ($938.75)

SOLUTION

a. For the J.C. Penney bond, we have $P = \$1000.00$, $r = 0.07375$, $p = \$1023.75$, and $N = 2004 - 1997 = 7$. Thus, the approximate yield to maturity is

$$R^* \approx \frac{NPr + P - p}{N\left(\frac{P + p}{2}\right)}$$

$$= \frac{(7)(\$1000.00)(0.07375) + \$1000.00 - \$1023.75}{(7)\left(\frac{\$1000.00 + \$1023.75}{2}\right)}$$

$$\approx 0.070$$

$$= 7.0\%.$$

b. For the Tenneco bond, we have $P = \$1000.00$, $r = 0.0725$, $p = \$938.75$, and $N = 2025 - 1997 = 28$. Thus, the approximate yield to maturity is

$$R^* \approx \frac{NPr + P - p}{N\left(\frac{P + p}{2}\right)}$$

$$= \frac{(28)(\$1000.00)(0.0725) + \$1000.00 - \$938.75}{(28)\left(\frac{\$1000.00 + \$938.75}{2}\right)}$$

$$\approx 0.077$$

$$= 7.7\%. \qquad \blacklozenge$$

EXAMPLE 5a

Steps	Display
1000 [+]	
1023.75 [=]	2023.75
[÷] 2 [=]	1011.875
[×] 7 [=] [STO]	7083.125
7 [×] 1000 [×]	
0.07375 [=]	516.25
[+] 1000	
[−] 1023.75 [=]	492.50
[÷] [RCL] [=]	0.069531457

CORPORATE BONDS SECTION 3

Try one!

Approximate the yield to maturity for the following bonds purchased in January 1997. Use the information given on Table 5.

a. Data General

Answer: _____

b. Proctor and Gamble

Answer: _____

As seen in Example 5, the current price for a bond is usually such that its yield to maturity approximates the annual percentage rate of newly issued bonds. (In 1997, the annual rate of newly issued bonds was about 7.0%.) Table 6 lists the yield to maturity for the bonds given in Tables 4 and 5.

TABLE 6

Yield to Maturity for Selected Corporate Bonds (January 1997)

Issuer	Rate r	Par Value P	Years to Maturity, M	Current Price P	Yield to Maturity, R^*
Cambridge Electric Light	$7\frac{3}{4}\%$	$1000.00	5	$1002.50	7.7%
Data General	$8\frac{3}{8}\%$	$1000.00	5	$995.00	8.5%
Exxon Corporation	$8\frac{1}{4}\%$	$1000.00	2	$1050.00	5.6%
Florida Power and Light	9.8%	$1000.00	21	$1085.00	9.0%
Ford Motor Company	9.15%	$1000.00	7	$1012.50	8.9%
General Electric	$7\frac{7}{8}$	$1000.00	1	$1028.75	4.9%
J.C. Penney	$7\frac{3}{8}\%$	$1000.00	7	$1023.75	7.0%
Owens-Corning	5.79%	$1000.00	8	$980.00	6.1%
Proctor and Gamble	8.0%	$1000.00	32	$1100.00	7.3%
Tenneco, Inc.	$7\frac{1}{4}\%$	$1000.00	28	$938.75	7.7%

As with stock prices, daily quotations for bond prices are listed in the financial section of most major newspapers. Table 7 shows a representative listing of bonds traded on the New York Exchange on February 13, 1997.

TABLE 7

New York Exchange Bonds: Daily Quotations, February 13, 1997

Bonds	Current Yield	Volume	Close	Net Change
Florsh $12\frac{3}{4}02$	11.8	140	$107\frac{5}{8}$	$+\ \frac{1}{2}$
FordCr $6\frac{3}{8}08$	6.7	16	$94\frac{7}{8}$	$+\ \frac{1}{8}$
GPA Del $8\frac{3}{4}98$	8.6	2	$101\frac{1}{2}$	...
GnCorp 8s02	6.6	60	122	$+\ 3$
GHost $11\frac{1}{2}02$	11.6	256	$99\frac{1}{2}$	...
GHost 8s02	cv	60	82	$-\ \frac{1}{2}$
GMA $7\frac{3}{4}97$	7.7	10	$100\frac{3}{16}$	$+\ \frac{1}{32}$
GMA $8\frac{3}{8}97$	8.4	50	$100\frac{3}{16}$	...
GMA 8.40s99	8.0	10	$104\frac{5}{8}$	...

Note from Table 7 that a given corporation may offer several different types of bonds. The listing immediately following the corporation name represents the coupon rate, the number of times interest is paid per year (s for semiannual, no letter for annual), and the year of maturity. For example, the first bond listed in Table 7 has an annual percentage rate of $12\frac{3}{4}\%$, the interest is paid annually, and the maturity date is 2002. A "cv" listing in the current yield column means that the bond is **convertible** to common stock. For example, the General Host (8s02) convertible bond listed in Table 7 has the option to convert General Host bonds into shares of common stock at a price established at bond issue, regardless of the current price of these common stocks. The value of this particular bond feature at any particular time is determined by the current stock price. When the conversion price for the stock is considerably less than the current price for the stock, this feature is very valuable. However, when the bond conversion price is greater than the current price of the stock, this feature is of little value.

When interest is paid on a bond, the entire amount of interest is sent to the current **bondholder**. Thus, when a bondholder sells a bond on a day other than an interest payment day, the new bond holder must reimburse the old bondholder for **interest accrued** since the last interest payment date. For example, if a bond sale were settled 45 days after its most recent interest payment date, the bond purchaser must include 45 days worth of interest in the purchase price.

A **stockbroker's commission** is also included in the purchase price of a bond. As with stocks, this commission varies from one broker to another. We call the sum of the current price, accrued interest, and broker's commission the **total purchase price** of a bond.

$$\text{Total Purchase Price} = \text{Current Price} + \text{Accrued Interest} + \text{Broker's Commission}$$

EXAMPLE 6 Finding the total purchase price of a bond

Ellen Luther purchased ten bonds (5.3s02) for the closing price on August 17, 1979, which was $82\frac{7}{8}$. The interest payment dates for these bonds are May 1 and November 1. The settlement date for the transfer of ownership of the bonds was August 23, 1979 (four business days after the day of purchase). If Ellen paid a broker's commission of $10.00 per bond plus accrued interest up to (but not including) the settlement date, what was the total purchase price? (Assume the par value is $1000.00.)

SOLUTION

The current price is given as $82\frac{7}{8}\%$ of $1000.00, or $828.75. The number of days from May 1 to August 23 is

$$31 + 30 + 31 + 22 = 114 \text{ days.}$$

This means that the accrued interest for each bond is

$$I = PrT = (\$1000.00)(0.053)\left(\tfrac{114}{365}\right) \approx \$16.55.$$

(Remember that when calculating interest, T stands for the amount of *time in years*. Since we are dealing with 114 days, we need to find out the fraction of a year represented by 114 days.)

Adding $10.00 broker's commission, we have a total purchase price for each bond amounting to

$$\text{Total Purchase Price} = \text{Current Price} + \text{Accrued Interest} + \text{Broker's Commission}$$

$$= \$828.75 + \$16.55 + \$10.00$$

$$= \$855.30.$$

Thus, Ellen's cost for the ten bonds was

$$(10)(\$855.30) = \$8553.00. \qquad \blacklozenge$$

CORPORATE BONDS

Try one!

Ronald Jones purchased eight electric company bonds ($7\frac{7}{8}$s99) for the closing price February 3, 1997 which was 101. Assume that the interest payment dates for these bonds are June 1 and December 1. The settlement date for the transfer of ownership of the bonds was February 6 (three business days after the day of purchase). If Ronald paid a broker's commission of $10.00 per bond plus accrued interest up to (but not including) the settlement date, what was the total purchase price? (Assume the par value is $1000.00.)

Answer: _____

When an individual buys and sells bonds, the net income realized is affected in at least three different ways: the amount of interest paid to the bondholder while the bond is held, the broker's commissions paid for buying and selling, and the difference in the purchase and selling prices of the bonds. We call the actual realized rate of earnings for the purchase and sale of bonds the **annual effective yield**.

$$\text{Annual Effective Yield} = \frac{\text{Interest Earned}}{(\text{Purchase Price})(\text{Time in Years})}$$

The following example illustrates how to calculate the net income and the annual effective yield for bonds.

EXAMPLE 7 Finding the net income and annual effective yield on the purchase and sale of bonds

William Brennan bought 20 convertible bonds (5s98) on January 15, 1995 at 73 each. On August 6, 1995, William sold the bonds at $78\frac{3}{8}$. If the broker's commission was $2.50 per bond for buying and again for selling, what was William's net income and what was his annual effective yield?

SOLUTION

The number of days from January 15 to August 6 is calculated below.

17	January
28	February
31	March
30	April
31	May
30	June
31	July
5	August
203	

At 5%, the interest for 203 days would amount to

$$I = PrT = (\$1000.00)(0.05)\left(\tfrac{203}{365}\right) \approx \$27.81 \text{ per bond.}$$

Thus, the net income on William's investment is

$15,675.00	selling (ending) price:	(20)($1000.00)(0.78375)
− 14,600.00	purchase (beginning) price:	(20)($1000.00)(0.73)
− 50.00	selling commission:	(20)($2.50)
− 50.00	purchase commission:	(20)($2.50)
+ 556.20	interest:	(20)($27.81)
$1,531.20	net income	

A net income of $1531.20 on an investment of $14,650.00 (purchase price plus commission) for 203 days corresponds to an annual effective yield Y of

$$Y = \frac{\$1{,}531.20}{(\$14{,}650.00)\left(\tfrac{203}{365}\right)} \approx 0.188 = 18.8\%. \qquad \blacklozenge$$

EXAMPLE 7

Steps	Display
203 ÷ 365 =	0.556164384
× 14650 =	8147.808219
1/x × 1531.2 =	0.18792784

CORPORATE BONDS   SECTION 3 **75**

Note in Example 7 that it is not necessary to know the interest payment dates to find the net (or total) income. We are simply concerned that the bondholder ends up with interest corresponding to the total number of days the bond was held. However, be aware that the seller of the bond would have already received one coupon payment for that year and would only be paid the additional amount of interest making up the difference between that coupon payment and the total amount of interest owed.

Try one!

Andrea Willis bought ten convertible bonds (6.5s02) on March 10, 1995 at 75 each. On October 8, 1995, Andrea sold the bonds at $79\frac{5}{8}$. If the broker's commission was $3.00 per bond for buying and again for selling, what was Andrea's net income and what was her annual effective yield?

Net Income: _____   Annual Effective Yield: _____

Important Terms

accrued interest

annual effective yield

bondholder

broker's commission

convertible bond

corporate bond

debenture bond

issuer of bond

maturity date

nondebenture bond

N, years to maturity

P, par value

p, current price

r, annual coupon rate

R, current yield

R^*, yield to maturity

total income

total purchase price

Important Formulas

Total Income = Selling Price + Interest − Purchase Price

$$\text{Current Yield} = R = \frac{\text{Annual Interest}}{\text{Current Price}} = \frac{Pr}{p}$$

$$\text{Yield to Maturity} = R^* \approx \frac{NPr + P - p}{N\left(\frac{P + p}{2}\right)}$$

Total Purchase Price = Current Price + Accrued Interest + Broker's Commission.

$$\text{Annual Effective Yield} = \frac{\text{Interest Earned}}{(\text{Purchase Price})(\text{Time in Years})}$$

SECTION 3 EXERCISES

1. Chuck and Mary Carr purchased five newly issued General Telephone Company of Michigan bonds (par value $1000.00) on September 1, 1976 at 100. The coupon rate is $8\frac{1}{2}\%$ with interest payable March 1 and September 1 and the maturity date is September 1, 2006.

   a. Find the term of the bonds.

   Answer: _____

   b. Find the purchase price of the five bonds. (Ignore broker's fees.)

   Answer: _____

   c. How much interest will Chuck and Mary receive from these bonds every six months?

   Answer: _____

   d. Find the total income if the bonds are held to maturity.

   Answer: _____

   e. Find the total income if Chuck and Mary sold the bonds at 82 on September 1, 1979.

   Answer: _____

2. Fred and Patty Clark purchased ten newly issued corporate bonds (par value $1000.00) on September 15, 1989 at 100. The coupon rate is 7.6% with interest payable March 15 and September 15 and the maturity date is September 15, 2014.

   a. Find the term of the bonds.

   Answer: _____

   b. Find the purchase price of the ten bonds. (Ignore broker's fees.)

   Answer: _____

   c. How much interest did Fred and Patty receive every six months?

   Answer: _____

   b. Find the total income if the bonds were held to maturity. (Ignore broker's fees.)

   Answer: _____

   e. Find the total income if Fred and Patty sell the bonds at $86\frac{3}{4}$ on September 15, 1997.

   Answer: _____

CORPORATE BONDS

3. Walter Johnson purchased the bonds sold by the Carrs in part **e** of Exercise 1.

   a. Find the current yield on this investment.

   Answer: _____

   b. Find the total income if Walter holds the bonds to maturity. (Ignore broker's fees.)

   Answer: _____

   c. Find the approximate yield to maturity.

   Answer: _____

4. Jean Ann Rogers purchased the bonds sold by the Clarks in part **e** of Exercise 2.

   a. Find the current yield on this investment.

   Answer: _____

   b. Find the total income if Jean Ann holds the bonds to maturity. (Ignore broker's fees.)

   Answer: _____

   c. Find the approximate yield to maturity.

   Answer: _____

5. Terry Foster purchased three newly issued corporate bonds (par value $1000.00) on September 1, 1991 at $99\frac{3}{8}$. The annual percentage rate is 10.75% with interest payable March 1 and September 1 and the maturity date is September 1, 2021.

   a. Find the term of the bonds.

   Answer: _____

   b. Find the purchase price of the three bonds. (Ignore broker's fees.)

   Answer: _____

   c. Find the total income if the bonds are held to maturity.

   Answer: _____

   d. Find the total income if Terry sold the bonds at $99\frac{1}{4}$ on September 1, 1996.

   Answer: _____

CORPORATE BONDS

6. Carla Dunham purchased five newly issued electric company bonds (par value $1000.00) on September 1, 1990 at $100\frac{3}{8}$. The annual percentage rate is $7\frac{1}{2}\%$ with interest payable March 1 and September 1 and the maturity date is September 1, 2020.

   a. Find the term of the bonds.

   Answer: _____

   b. Find the purchase price of the five bonds. (Ignore broker's fees.)

   Answer: _____

   c. Find the total income if the bonds are held to maturity.

   Answer: _____

   d. Find the total income if Carla sells the bonds at 78 on September 1, 1998.

   Answer: _____

7. Douglas Powers purchased the bonds sold by Terry Foster in part **d** of Exercise 5.

   a. Find the current yield on this investment.

   Answer: _____

   b. Find the total income if Douglas holds the bonds to maturity.

   Answer: _____

   c. Find the approximate yield to maturity.

   Answer: _____

8. Janet Mace purchased the bonds sold by Carla Dunham in part **d** of Exercise 6.

   a. Find the current yield on this investment.

   Answer: _____

   b. Find the total income if Janet holds the bonds to maturity.

   Answer: _____

   c. Find the approximate yield to maturity.

   Answer: _____

CORPORATE BONDS

9. Complete the following table by finding the current yield and the yield to maturity.

Current Price	Coupon Rate	Years to Maturity	Current Yield	Yield to Maturity
a. $99\frac{1}{8}$	$10\frac{1}{2}$	30	_____	_____
b. $86\frac{5}{8}$	$6\frac{7}{8}$	13	_____	_____
c. 53	$4\frac{5}{8}$	15	_____	_____
d. $103\frac{1}{2}$	$10\frac{1}{4}$	30	_____	_____

10. Complete the following table by finding the current yield and the yield to maturity.

Current Price	Coupon Rate	Years to Maturity	Current Yield	Yield to Maturity
a. $107\frac{1}{4}$	$11\frac{1}{4}$	25	_____	_____
b. 60	$4\frac{3}{4}$	11	_____	_____
c. $79\frac{1}{2}$	$4\frac{1}{4}$	4	_____	_____
d. $56\frac{3}{4}$	7	15	_____	_____

The bonds in Exercises 11–14 are from Table 7 on page 70.

11. What is the annual coupon rate and the year of maturity for the General Host (8s02) bonds?

Annual Coupon Rate: _____   Year of Maturity: _____

12. What is the annual coupon rate and the year of maturity for the General Host ($11\frac{1}{2}$02) bonds?

Annual Coupon Rate: _____   Year of Maturity: _____

13. What was the closing price (see Table 7) of the GnCorp (8s02) bonds on

   a. February 13, 1997?

                                                      Answer: _____

   b. February 12, 1997?

                                                      Answer: _____

14. What was the closing price (see Table 7) of the GMA (8.40s99) on

   a. February 13, 1997?

                                                     Answer: _____

   b. February 12, 1997?

                                                     Answer: _____

CORPORATE BONDS   SECTION 3   **85**

15. Greg Van Campen purchased five telephone company (8.80s05) bonds with interest payable May 15 and November 15. He purchased the bonds at $92\frac{3}{8}$ (the closing price on September 10, 1989) and the settlement date was September 14, 1989. The commission for Greg's broker was $10.00 per bond.

  a. What was the accrued interest on each bond?

                                                                    Answer: _____

  b. What was the total purchase price for the five bonds?

                                                                    Answer: _____

16. Allison Thornton purchased five automobile corporation (8.75s00) bonds with interest payable February 1 and August 1. She purchased the bonds at $92\frac{1}{4}$ on September 10, 1990 and the settlement date was September 14, 1990. The commission was $10.00 per bond.

  a. What was the accrued interest on each bond?

                                                                    Answer: _____

  b. What was the total purchase price for the five bonds?

                                                                    Answer: _____

86 SECTION 3   CORPORATE BONDS

17. On June 4, 1974, Larry Rhodes bought 15 Tenneco corporation ($6\frac{1}{4}92$) convertible bonds at $83\frac{1}{2}$. On September 10, 1979, he sold the bonds for 135 each. If the broker's commission was $5.00 per bond for buying and again for selling, what was Larry's net income and his annual effective yield? (Larry held these bonds for a total of 1924 days.)

Net Income: _____

Annual Effective Yield: _____

18. On June 4, 1984, Wendy and John Fischer bought ten shares of a furniture company's convertible bonds ($5\frac{3}{4}04$) at 61. On September 10, 1989, they sold the bonds at 85. If the broker's commission was $7.50 per bond for buying and again for selling, what was the Fischer's net income and their annual effective yield? (The Fischers held these bonds for a total of 1,924 days.)

Net Income: _____

Annual Effective Yield: _____

Section 4
Mutual Funds

Closely monitoring the stock and bond markets can be a time consuming endeavor. Partly for this reason, many people choose to invest in mutual funds. A mutual fund is a firm that manages its clients' money by investing it in a variety of ways. Some mutual funds invest in stocks, some invest in bonds, while still others invest in a combination of the two. A stock fund typically has millions of dollars in assets and purchases stock from a variety of companies. For this reason, the performance of any one company will have little impact on the value of the fund. Because of this asset diversification, investment in a stock fund is less risky than investment in the stock of a single company. Bond funds offer similar advantages.

Mutual funds are often classified according to their investment objective. These include growth, equity income, growth and income, and bond funds. In addition, international funds invest in the stocks and bonds of companies in foreign countries.

A **growth fund** invests primarily in companies that have a great potential for future growth, as reflected in a high price-earnings ratio. These companies generally reinvest their profits and pay little or no dividends. Therefore, investors in a growth fund look to benefit from the rise in value of the stocks owned by the fund, rather than the dividends paid out.

A **growth and income fund** invests in companies with significant growth potential and that pay moderate dividends, as well as bonds. These funds combine the potential for significant increases in the value of the underlying stocks with some periodic income from dividends and interest.

An **equity income fund** invests in the stock of companies that pay out large dividends, and in bonds. Dividends from the stocks and interest from the bonds are paid out periodically to the mutual fund shareholders, thus providing them with a consistent income.

A **bond fund** invests in corporate and government bonds, and typically provides a steadier income to the shareholders than do stock funds. However, over the long term, stock funds generally provide more income than do the more conservative bond funds.

Table 8 illustrates a typical newspaper listing of mutual fund activity.

TABLE 8

Wall Street Journal Mutual Funds Quotations, February 28, 1996

Fund Name	NAV (Net asset value)	Net Change	YTD % Ret (Year-to-date percent return)
Riverside Capital:			
Equity	14.13	–0.01	+5.6
Fxdin	9.13	–0.02	–1.5
Grow	13.16	–0.02	+2.9
TNMuOb	9.75	...	–0.1
Robertson Stephens:			
Contra	15.99	+0.13	+16.0
Dev Ctry	8.79	+0.04	+9.6
Em Gr	19.62	–0.02	+2.1
Gr Inc	11.99	–0.06	+6.7
Rodney Square:			
Divin	13.07	–0.03	–0.4
Gwth	18.31	–0.05	+5.7
IntlEq	12.51	...	+3.2

Column one lists the names of the mutual funds and the companies that manage them, also know as the *fund family*. Column two gives the prior day's closing price per share (February 27, 1996). Column three provides the daily change in the share price, and column four lists the percent change in the fund's share price since January 1, 1996.

A fund's share price is called the **net asset value** (*NAV*). This is computed as follows.

$$NAV = \frac{\text{Total Net Assets of Fund}}{\text{Number of Shares Outstanding}}$$

$$= \frac{\text{Fund Assets} - \text{Fund Liabilities}}{\text{Number of Shares Outstanding}}$$

The fund assets include, among other things, the stocks and bonds owned by the fund. Liabilities include, among other things, fees owed by the fund to stock brokers, with whom the fund managers interact, and fees for the management of the fund.

EXAMPLE 1 Reading a table of mutual funds quotations

Use Table 8 to answer the following questions about mutual fund prices.

a. Determine the net asset value (*NAV*) of Robertson Stephens Emerging Growth Fund at the close of the day on February 27, 1996.

b. Determine the net asset value (*NAV*) of Robertson Stephens Emerging Growth Fund at the close of the day on February 26, 1996.

SOLUTION

a. Table 8 lists the closing prices for the previous day. Therefore, the net asset value (*NAV*) of the Robertson Stephens Emerging Growth Fund at the close of the day on February 27 was $19.62.

b. The net change for this fund was −0.02. This means that the net asset value (*NAV*) fell $0.02 from the close of the previous day. Therefore,

Previous Day's Closing Price = *NAV* − Net Change

$$= \$19.62 - (-\$0.02)$$

$$= \$19.62 + \$0.02$$

$$= \$19.64.$$ ◆

Try one!

Use Table 8 to answer the following questions about mutual fund prices.

a. Determine the net asset value (*NAV*) of Riverside Capital Equity Fund at the close of the day on February 27, 1996.

Answer: _____

b. Determine the net asset value (*NAV*) of Riverside Capital Equity Fund at the close of the day on February 26, 1996.

Answer: _____

Every Friday the *Wall Street Journal* provides an expanded table of mutual funds quotations. In addition to the information shown in Table 8, an expanded table includes four-week, one-, three-, and five-year total returns, and fund expenses. It also gives the investment objective and a performance ranking for each fund listed. Selected entries are given in Table 9.

MUTUAL FUNDS

TABLE 9

Wall Street Journal Mutual Funds Quotations, February 28, 1997

NAV	Net Chg	Fund Name	Inv Obj	YTD %ret	4Wk %ret	1Yr	Total Return 3Yr-R	5Yr-R	Max Init Chrg	Exp Ratio
		Evergreen Funds:								
20.05	-0.43	AggGroA	CP	-1.8	-5.2	+9.6	+11.2	+12.8	4.75	1.22
13.34	-0.10	BalA	MP	+3.0	+1.4	+12.4	+12.1	+11.2	4.75	0.88
		Fidelity Invest:								
45.38	-0.23	EqInc	EI	+6.0	+3.2	+22.4	+18.9	+17.8	0.00	0.68
24.80	-0.19	EQII	EI	+4.9	+2.9	+18.8	+16.6	+16.7	0.00	0.73
14.43	-0.02	Nordic	IL	+4.6	+4.3	+36.6	NS	NS	3.00	2.00
		Price Funds:								
20.04	-0.19	BlChip	GR	+5.1	+2.2	+25.7	+22.7	NS	0.00	1.25
13.94	...	IntlStk	IL	+1.0	+3.2	+12.5	+8.3	+11.5	0.00	0.88

EXAMPLE 2   **Reading an expanded table of mutual funds quotations**

Jim and Nancy Matthews purchased 100 shares of Fidelity Equity Income on February 27, 1996 with a net asset value (*NAV*) of $39.75. Use Table 9 to answer the following questions.

a. What was the net asset value (*NAV*) on February 27, 1997?

b. What was the total value of Jim and Nancy's 100 shares on February 27, 1997?

c. What was the total value of Jim and Nancy's 100 shares on February 27, 1996?

d. Has the total value of the 100 shares increased or decreased and by how much?

SOLUTION

a. From the first column in Table 9 we see that the net asset value for one share of Fidelity Equity Income was $45.38.

b. The total value of Jim and Nancy's 100 shares on February 27, 1997 was

Total Value = (number of shares)(net asset value)

= (100)($45.38)

= $4538.00.

c. Since the net asset value on February 27, 1996 was $39.75, the total value on this date was

Total Value = (100)($39.75)

= $3975.00.

d. The total value for the 100 shares increased from $3975.00 to $4538.00. The amount of the increase was

$4538.00 − $3975.00 = $563.00. ◆

Note that Example 2 does not consider any dividends that Jim and Nancy may have received from the fund during the year. These dividends would increase the attractiveness of investment in this fund.

Try one!

Bob and Stacey Carlson purchased 200 shares of Evergreen Aggressive Growth A on February 27, 1996 with a net asset value of $18.30. Use Table 9 to determine the amount of increase in the total value of their 200 shares between February 27, 1996 and February 27, 1997.

Answer: _____

The best measure of a mutual fund's performance is the **total return**. The total return is the sum of net asset value (*NAV*) increases or decreases plus any capital gains or dividends that were paid during a particular time period (such as four weeks, year-to-date, one year, three years, or five years) divided by the net asset value at the beginning of the time period. The total return for a fund is computed by the mutual fund company and is listed in the expanded mutual funds quotations as shown in Table 9.

The one-year total return can be used to determine the net income an individual would achieve if mutual fund shares were bought at the start and sold at the end of the one-year period. An individual's one year net income is

Net Income = (One-Year Total Return)(P)(N)

where

P = purchase price per share

and

N = number of shares purchased.

EXAMPLE 3 Calculating the one-year net income

Ron and Cindy Barnett bought 200 shares of Fidelity Equity Income II at $22.37 per share on February 27, 1996. Use Table 9 to answer the following.

a. What was the one-year total return on the Barnetts' investment?

b. What was their one-year net income?

SOLUTION

a. According to Table 9, the one-year total return for Fidelity Equity Income II on February 27, 1997 was 18.8%.

b. Since the one-year total return was 18.8% = 0.188 and 200 shares were purchased at $22.37 per share, we have

Net Income = (0.188)($22.37)(200) ≈ $841.11.

Try one!

James and Holly Kramer bought 150 shares of Price Blue Chip at $16.12 per share on February 27, 1996. Use Table 9 to answer the following.

a. What was the one-year total return on the Kramers' investment?

Answer: _____

b. What was their one-year net income?

Answer: _____

For some funds a load is charged. A **load** is a fee paid by the investor to the mutual fund company. A load is similar to a stock broker's commission in that it is a percentage of the total purchase price of the investment. Note that Table 9 includes a column listing the load amount for each fund as a percent (Max Init Chrg). When no load is charged, the fund is called a **no-load fund**.

EXAMPLE 4　Calculating the load on a mutual fund purchase

Debbie and John Miller bought 100 shares of Evergreen Aggressive Growth A on February 27, 1997 at the closing price. Determine the purchase price for the shares and the dollar amount of the load.

SOLUTION

From Table 9, we see that the net asset value for Evergreen Aggressive Growth A on February 27, 1997 was $20.05. Therefore, the purchase price for the 100 shares was

$(100)(\$20.05) = \$2005.00.$

From Table 9, we see that the load was 4.75%, so the dollar amount of the load is

$(0.0475)(\$2005.00) \approx \$95.24.$ ◆

Note that in Example 4 the Millers would pay the mutual fund company a total of $2005.00 + $95.24 = $2100.24 to purchase these shares.

Try one!

What was the dollar amount of the load for the purchase of 250 shares of Fidelity Nordic at the closing price on February 27, 1997?

Answer: _____

One major benefit of investment in a mutual fund is that a staff of professional fund managers is at work analyzing the market in an attempt to maximize the return on the shareholders' investment. Although the individual shareholder has no say in the allocation of assets within a fund, it is the responsibility of the fund to periodically notify its shareholders of its activities and performance. The expenses incurred in the management of the fund, as well as the costs of providing shareholders with periodic financial reports, are paid out of the fund's assets. These expenses are listed as the **expense ratio** in Table 9. The expense ratio is computed as follows.

$$\text{Expense Ratio} = \frac{\text{Fund Management Expenses}}{\text{Total Fund Assets}}$$

The expense ratio is typically reported as a percent.

Management expenses can have a significant impact on an individual's return from a mutual fund investment. The **gross return before expenses** reflects the fund's one-year total return before management expenses are subtracted. Thus, it shows the hypothetical one-year return had no management expenses been incurred. The formula for gross return is given below.

$$\text{Gross Return Before Expenses} = \text{One-Year Total Return} + \text{Expense Ratio}$$

EXAMPLE 5 Calculating the gross return before expenses

Glenn and Mary Morgan purchased some shares of Evergreen Aggressive Growth A on February 27, 1996. Use Table 9 to answer the following questions.

a. What was their one-year total return?

b. What was their gross return before expenses?

SOLUTION

a. According to Table 9, the one-year total return for Evergreen Aggressive Growth A on February 27, 1997 was 9.6%.

b. Since the one-year total return was 9.6%, and the expense ratio, listed in Table 10, was 1.22%, we have

Gross Return Before Expenses = 9.6% + 1.22%

= 10.82%.   ◆

In Example 5, Glenn and Mary's one-year return on their investment was significantly reduced by mutual fund management expenses.

MUTUAL FUNDS   SECTION 4   **97**

Try one!

Roger and Debbie Hayes purchased shares of Fidelity Nordic on February 27, 1996. Use Table 9 to answer the following questions.

a. What was their one-year total return?

Answer: _____

b. What was their gross return before expenses?

Answer: _____

In order to minimize management expenses, some investors choose to purchase shares in **index funds**. An index fund buys stock from companies listed in a specific stock index. Examples of stock indexes are *Standard and Poor's 500 Index*, which includes 500 medium to large firms, and the *NASDAQ Composite Index* of nearly 3000 smaller firms. Because an index fund buys stock from each company in a specific index, no research into the investment potential of the stocks is required. Therefore, management expenses of index funds are typically much lower than those of other stock funds. In addition to index stock funds, there are **index bond funds**, offering similar advantages.

Mutual funds may be purchased directly or through a stock broker. There are at present hundreds of mutual funds, some with loads and some without, some with high management expenses and some with low management expenses, some domestic and some international. This variety gives the investor great flexiblity but can also make it difficult to decide which funds to purchase. The local library has information on the present and past performance of the different mutual funds.

Important Terms

bond fund	gross return before expenses
equity income fund	N, number of shares purchased
expense ratio	net asset value (NAV)
growth fund	net income
growth and income fund	no-load fund
index bond funds	P, purchase price per share
index funds	total return
load	

Important Formulas

$$NAV = \frac{\text{Total Net Assets of Fund}}{\text{Number of Shares Outstanding}}$$

Net Income = (One-Year Total Return)$(P)(N)$

Total Value = (number of shares)(NAV)

$$\text{Expense Ratio} = \frac{\text{Fund Management Expenses}}{\text{Total Fund Assets}}$$

Gross Return Before Expenses = One–Year Total Return + Expense Ratio

MUTUAL FUNDS SECTION 4

SECTION 4 EXERCISES

1. Use Table 8 on page 88 to determine the Net Asset Value (*NAV*) of Rodney Square Growth Fund at the close of the day on February 27, 1996.

Answer: _____

2. Use Table 8 to determine the *NAV* of Rodney Square Growth Fund at the close of the day on February 26, 1996.

Answer: _____

3. Of the funds listed in Table 8, which performed the best between January 1, 1996 and February 27, 1996? Which performed the worst?

Best: _____   Worst: _____

4. Bob and Susan Woodson purchased 200 shares of Price International stock on February 27, 1996 with an *NAV* of $12.74.

   a. What was the total value of Bob and Susan's 200 shares on the date the shares were purchased?

   Answer: _____

   b. Use Table 9 on page 91 to find the *NAV* and the total value of Bob and Susan's 200 shares of Price International stock on February 27, 1997.

   NAV: _____   Total Value: _____

   c. By how much did the total value of the 200 shares increase or decrease over the year?

   Answer: _____

MUTUAL FUNDS

5. Sam and Tammy Cook bought 150 shares of Evergreen Balanced A Fund at $13.40 per share on February 27, 1996. Use Table 9 to answer the following questions.

   a. What was the one-year total return on the Cooks' investment?

   Answer: _____

   b. What was their one-year net income from this investment?

   Answer: _____

6. Stuart Thompson purchased 125 shares of Fidelity Equity Income with an NAV of $39.75 on February 27, 1996.

   a. How much did Stuart pay for the 125 shares?

   Answer: _____

   b. What was the one-year total return on Stuart's investment?

   Answer: _____

   c. What was the one-year net income from this investment?

   Answer: _____

102 SECTION 4   MUTUAL FUNDS

7. Ron and Sandy Taylor purchased 275 shares of Evergreen Balanced A at the closing price on February 27, 1997. Use Table 9 to answer the following questions.

   a. What was the purchase price of the 275 shares and the dollar amount of the load?

   Purchase Price: _____   Load Amount: _____

   b. With the load, how much did Ron and Sandy pay for the 275 shares?

   Answer: _____

8. Walter and Sara Johnson purchased 100 shares of Fidelity Nordic at the close of the day on February 27, 1997. Steve and Mary Watson purchased 200 shares of Fidelity Nordic on the same day at the same price. Use Table 9 to answer the following questions.

   a. What was the dollar amount of the load paid by the Johnsons for their 100 shares?

   Answer: _____

   b. What was the dollar amount of the load paid by the Watsons for their 200 shares?

   Answer: _____

MUTUAL FUNDS   SECTION 4 **103**

9. Using Table 9, which of the funds listed are no-load funds?

Answer: _____

_____

_____

10. Sam and Jerri Baker purchased shares of Fidelity Equity Income II on February 27, 1996. Use Table 9 to determine the one-year total return of this investment, and the one-year gross return before expenses.

One Year Total Return: _____

Gross Return: _____

Section 5
Spotlight on the Financial Manager

If you are interested in a career involving finance, nearly every type of business has a place for you whether it be in education, communications, manufacturing, health care, banking, or just about any field you can imagine. Every business needs capital to operate, and it is the job of the financial managers to see that this capital is handled properly. The positions available for financial managers range from that of controller and treasurer to credit and cash manager, and even all the way to chief financial officer (CFO) of a corporation.

Controllers are in charge of reporting the financial status of the company through the use of reports and statements. They are usually responsible for balancing the books and overseeing the accounting, auditing, and budget departments. Cash and credit managers handle the day-to-day cash flow. They manage the business and investment needs of the firm. It is the cash and credit managers that determine if there is a cash deficit or surplus, and proceed accordingly by either obtaining a loan, or seeking out interest-earning investment opportunities. Bank financial managers work as loan officers, investment counselors, and credit or trust managers. They may work in a savings and loan, consumer credit institution, or a credit union.

A minimum of a bachelor's degree in accounting, finance, or business administration is required to prepare for a career as a financial manager. If you plan a career in international finance, fluency in at least one foreign language is desirable. Some financial managers gain additional education and training by obtaining certification from a professional association such as the *Association for Investment Management and Reasearch*. This organization awards the title of "Chartered Financial Analyst" to those who pass three exam levels and have three years of work experience. Another such organization is *The National Association of Credit Management* whose certification program has three levels: Credit Business Associate, Credit Business Fellow, and finally Certified Credit Executive.

Although your career in financial management may start out at a low level, do not be discouraged. Take advantage of the opportunity to learn the workings of the company, find a mentor, and invest in yourself. Patience is very important when your goal is advancement within a company.

Starting salaries for those in corporate finance range from $25,000 to $35,000 for those holding a bachelor's degree. Your title at the entry level with a bachelor's degree may be *junior financial analyst*. With a master's degree you may start with the title *financial analyst*. Financial analysts can earn from $23,000 to $47,000 depending on the firm's size and location and the analyst's particular skill level. Senior financial analysts typically make between $45,000 and $70,000. They usually have a master's degree and at least three years of work experience. Credit managers can earn from $30,000 to $63,000, assistant treasurers from $40,000 to $78,000.

The chief financial officer of a small firm may personally handle the day-to-day financial decision-making, while at a large firm the CFO most likely oversees a staff in the financial department of the firm. There is usually only one CFO in a company, so these jobs are a bit more difficult to secure and require a greater degree of education and experience than do other financial occupations.

It is common for a CFO to hold at least a Master of Business Administration (MBA) degree. The position of CFO is often filled by promoting a highly skilled professional from within the company's own accounting, insurance, or budget department. The typical experience for CFOs in leading large corporations is 15+ years of work experience and an MBA. The salary for a CFO in 1995 ranged from $60,000 in the smallest firms to $295,000, depending on the firm's location and the qualifications of the individual holding the position. It is not uncommon for CFOs to increase their earnings through participation in a corporate bonus program.

Financial mangers can expect to work in a comfortable environment, and usually have their own offices. They work 40 to 55 hours per week or more and are often required to attend the meetings of professional organizations. Some may even be required to travel for their firms. Financial managers need to be tactful, analytical, creative, have good judgment, be able to work independently, have excellent communication skills (both written and oral), and be adept at interpreting detailed information. They must also have good computer skills since a great deal of the work is done with the help of computer spreadsheets (like Excel or LOTUS), word processing software, presentation packages, and mainframes. Financial managers must also have good interpersonal skills, particularly if they manage a staff of employees. According to one major executive in the airline industry, the "biggest weakness is a lack of people skills." It is just as important to be able to work with people as it is to balance the books.

The field of financial management is likely to grow in the next ten years. The future looks good for this career choice. Global trade, as well as the changing federal and state laws increase the need for highly skilled financial managers. Those who keep up with the latest trends and changes will have the greatest opportunities for advancement within the field. Therefore, continuing education is of paramount importance to keep current. Some firms will help cover the costs of such continuing education. Financial managers must have the ability to think globally, formulate strategies, and be team players.

If you are interested in pursuing a career in financial management or one of its related fields, you can contact one or more of the following organizations for more information.

American Bankers Association
Center for Banking Information
1120 Connecticut Avenue, NW
Washington, DC 20036

National Association of Credit Management (NACM)
Credit Research Foundation
8815 Centre Park Drive
Columbia, MD 21045-2117

Treasury Management Association
7315 Wisconsin Avenue
Bethesda, MD 20814

Association for Investment Management and Research
5 Boar's Head Lane
P. O. Box 3668
Charlottesville, VA 22903

Healthcare Financial Management Association
Two Westbrook Corporate Center
Suite 700
Westchester, IL 60154

If you are interested in pursuing a summer internship in finance, you can refer your questions to the National Directory of Internships at the following address.

National Society for Internships and Experiential Education
3509 Haworth Drive
Suite 207
Raleigh, NC 27609
1-(919) 787-3263

SECTION 5 EXERCISES

In the following exercises, imagine that you are a financial manager for a company and it is your job to follow and report on the company's stock, bond and mutual fund investments as well as to distribute information to your fellow employees about the performance of your own company's stock.

1. Last fiscal year your corporation paid out a total common stock dividend of $75,450,000 on 39,562,000 shares selling for $42\frac{1}{8}$ per share. Determine the annual dividend per share and the dividend yield.

   Annual Dividend: _____

   Dividend Yield: _____

2. One of the companies in which your firm invests declares 25% of its net corporate earnings of $12,962,000 for common stock dividends on 5,231,000 shares selling at $51\frac{1}{4}$. Determine the annual dividend per share and the dividend yield.

   Annual Dividend: _____

   Dividend Yield: _____

3. If another of your firm's investment companies has earnings of $1,325,137,000 and the average price per share is $32\frac{5}{8}$, determine the earnings per share and the price-earnings ratio for this investment if there are 158,628,000 shares.

Earning Per Share: _____

P-E Ratio: _____

4. You have been asked by your supervisor to chart the progress of the company's investment in Lockheed Martin. Using the newspaper report found in Table 3 (page 34), determine the following information.

   a. What was the 52-week high price for this stock on January 8, 1997?

   Answer: _____

   b. What was the closing price for this stock on January 8, 1997?

   Answer: _____

   c. What was the closing price for this stock on January 7, 1997?

   Answer: _____

5. You have determined that the firm needs to sell some of its stock in order to have some working capital. If your firm purchased 500 shares of Liz Claiborne common stock in January of 1996 at 30 and sold all the shares of this stock one year later at the closing price on January 8, 1997, what was net income and net annual return from this investment? (Assume that your firm received exactly one year of dividend payments at $0.45 per share and that there was a $50.00 broker's commission paid at both the purchase and sale of this stock.)

Answer: _____

6. Your firm purchased 2500 shares of a technology stock at $32\frac{3}{16}$ in February of 1995. In October of that same year there was a three-for-two stock split and in March of 1996 there was a two-for-one stock split. The stock was then sold in April 1996 at 30. Determine the purchase price, selling price, and total income of this stock. (Assume there were no dividends paid during this time period and ignore any broker's fees.)

Purchase Price: _____

Selling Price: _____

Total Income: _____

7. In January 1997, while reviewing the investment portfolio for your firm, you notice that the company is holding a number of corporate bonds. One of the investments in this bond portfolio is 20 Proctor and Gamble bonds purchased in January of 1989.

    a. Using the information in Table 4 on page 62, determine the annual interest paid on these 20 bonds, then calculate the total interest earned since purchase. (Assume that the firm has received the full interest payment for each of the years.)

    Annual Interest: _____   Total Interest: _____

    b. Using Table 5 on page 67, determine the number of years to maturity.

    Answer: _____

    c. Using Table 5 on page 67, approximate the yield to maturity for these bonds.

    Answer: _____

8. One of the ways in which your firm diversifies its assets is through investment in mutual funds. What was the Net Asset Value (*NAV*) of Robertson Stephens Growth and Income at the close of the day on February 27, 1996? (Use Table 8 on page 90.)

    Answer: _____

9. Using Table 8 on page 90, determine the *NAV* of Robertson Stephens Growth and Income at the close of the day on February 26, 1996.

    Answer: _____

10. One of your company's mutual fund investments is in Price International stock fund. Your firm purchased 1500 shares on February 27, 1996 when the *NAV* was $12.74.

a. What was the total value of the firm's 1500 shares on the date the shares were purchased?

Answer: _____

b. Using Table 9 on page 93, find the *NAV* and the total value of the firm's 1500 shares of Price International stock on February 27, 1997.

NAV: _____

Total Value: _____

c. By how much did this investment increase or decrease during the year?

Answer: _____

Solutions to "Try one!" Exercises

SECTION 1

Page 3

To find the dividend per share for 1995, we use $N = 141{,}000{,}000$ and $D = \$190{,}350{,}000.00$. Therefore, the dividend per share is

$$d = \frac{D}{N} = \frac{\$190{,}350{,}000.00}{141{,}000{,}000} = \$1.35.$$

Since Robert owns 200 shares of this common stock, his annual dividend is

$$\text{annual dividend} = (200)(\$1.35) = \$270.00.$$

Robert's quarterly dividend is

$$\text{quarterly dividend} = \frac{\text{annual dividend}}{4} = \frac{\$270.00}{4} = \$67.50.$$

Page 5

Since Robert's 1995 dividend per share was $1.35, the dividend yield in 1995 was

$$\text{dividend yield} = \frac{d}{p} = \frac{\$1.35}{50\frac{5}{8}} = \frac{\$1.35}{\$50.625} \approx 0.027 = 2.7\%.$$

Page 7

The earnings per share for Lockheed Martin Corporation for 1995 were

$$e = \frac{E}{N} = \frac{\$124{,}539{,}480.00}{73{,}692{,}000} = \$1.69.$$

Page 8

The earnings per share for 1995 were

$$e = \frac{E}{N} = \frac{\$651{,}414{,}560.00}{198{,}602{,}000} = \$3.28.$$

Therefore, the price-earnings ratio for a price of $65\frac{5}{8}$ was

$$\text{price-earnings Ratio} = \frac{p}{e} = \frac{\$65.625}{\$3.28} \approx 20.0.$$

Page 12

a. From Table 1, we see that General Motors paid dividends of $0.80, $0.80, and $1.10 per share during 1993, 1994, and 1995, respectively. Since Steve owns 300 shares, his total dividends were

1993 dividends = (300)($0.80) = $240.00
1994 dividends = (300)($0.80) = $240.00
1995 dividends = (300)($1.10) = $330.00.

b. The dividend yield on October 24, 1995 was

$$\text{dividend yield} = \frac{\$1.10}{44\frac{5}{8}} = \frac{\$1.10}{\$44.625} \approx 0.025 \approx 2.5\%.$$

Page 15

div. per share = $2.15 + $2.27 + $2.38 + $2.48 + $2.58 + $1.45 = $13.31

Since Madeline had 500 shares, her total dividends during the six-year period were

dividends = (500)($13.31) = $6655.00.

Since Madeline paid 38 per share and sold at 17, we have

purchase price = (500)($38.00) = $19,000.00

selling price = (500)($17.00) = $8500.00.

Thus, Madeline's total income was

total income = $8500.00 − $19,000.00 + $6655.00

= −$3845.00.

SOLUTIONS TO "TRY ONE" EXERCISES

Page 18

Since George had 1000 shares, his total dividends during the year were

$$\text{dividends} = (100)(\$1.00) = \$100.00.$$

Since George paid $40.625 per share and the closing price for the year was $62.375, we have

$$\text{purchase price} = (100)(\$40.625) = \$4062.50$$

$$\text{selling price} = (100)(\$110.50) = \$6237.50$$

Therefore, George's annual return was

$$\text{annual return} = \frac{\$6237.50 - \$4062.50 + \$100.00}{\$4062.50}$$

$$= \frac{\$2{,}275.00}{\$4062.50} = 0.560 = 56.0\% \text{ (quite a gain!)}.$$

SECTION 2

Page 36

a. The lowest price per share for Loctite common stock during the 52 weeks prior to January 8, 1997 was $42\frac{1}{4}$, or $42.25.

b. There were $(100)(105) = 10{,}500$ shares of common stocks traded on January 8, 1997.

c. The closing price for Microtest common stock on January 8, 1997 was $9\frac{3}{4}$ or $9.75 per share.

d. Since there was no net change in the closing price on January 8, 1997, the closing price for Microtest common stock on January 7, 1997 was $9\frac{3}{4} - 0 = 9\frac{3}{4}$, or $9.75.

Page 39

purchase price $= (200)(\$29.25) = \$5850.00$

purchase commission $= (0.008)(\$5850.00) = \$46.80$

dividends $= (200)(\$0.24) = \$48.00$

selling price $= (200)(\$37.00) = \$7400.00$

selling commission $= (0.005)(\$7400.00) = \$37.00$

net income $= SP - PP + D -$ commissions $= \$1514.20$

net annual return $= \dfrac{\$1514.20}{\$5850.00} \approx 0.259 = 25.9\%$

Page 41

purchase price $= (100)(\$47.50) = \$4750.00$

purchase commission $= \$50.00$

dividends $= (100)(\$1.20) = \$120.00$

selling price $= (100)(\$60.75) = \$6075.00$

selling commission $= \$65.00$

net income $= SP - PP + D -$ commissions $= \$1330.00$

net annual return $= \dfrac{\$1330.00}{\$4750.00} = 0.280 = 28.0\%$

SOLUTIONS TO "TRY ONE" EXERCISES

Page 42

a. Bill's purchase price for the stock was $(200)(\$103.50) = \$20,700.00$

b. Since there was a two-for-one split, his 200 shares become $(200)(\frac{2}{1}) = 400$ shares. Therefore, the selling price was $(400)(\$95.75) = \$38,300.00$.

Page 44

a. The Williams' purchase price for this stock was $(5000)(\$16.50) = \$82,500.00$.

b. The 4% stock dividend yielded $5000 + (0.04)(5000) = 5200$ shares.
The 5% stock dividend yielded $5200 + (0.05)(5200) = 5460$ shares.
The 5% stock dividend yielded $5460 + (0.05)(5460) = 5733$ shares.
Therefore, the selling price for the stock was $(5733)(\$18.00) = \$103,194.00$.

Page 47

a. The purchase price for the stock was $(10,000)(\$16.50) = \$165,000.00$

b. Since the Williams invested only $(5000)(\$16.50) = \$82,500.00$ of their own cash, they had to borrow $\$165,000.00 - \$82,500.00 = \$82,500.00$.

c. At 5% annual percentage rate for three years, the Williams paid
$I = PrT = (\$82,500.00)(0.05)(3) = \$12,375.00$ in interest.

d. The 4% stock dividend yielded $10,000 + (0.04)(10,000) = 10,400$ shares
The 5% stock dividend yielded $10,400 + (0.05)(10,400) = 10,920$ shares
The 5% stock dividend yielded $10,920 + (0.05)(10,920) = 11,466$ shares
Therefore, the selling price for the stock was $(11,466)(\$18.00) = \$206,388.00$

e. The net income (on margin) on this investment was

Net Income $= \$206,388.00 - \$165,000.00 + \$1000.00$
$- \$1856.94 - \$12,375.00 = \$28,156.06$.

SECTION 3

Page 61

For each bond, the interest for six months is

$$\text{interest} = PrT = (\$1{,}000.00)(0.095)\left(\tfrac{1}{2}\right) = \$47.50.$$

Since Steven and Ruby own seven bonds, their six-month interest checks will amount to

$$\text{total interest} = (7)(\$47.50) = \$332.50.$$

Page 64

Albert's interest per bond over the 25 years amounted to

$$\text{interest} = PrT = (\$1000.00)(0.0775)(25) = \$1937.50.$$

Thus, for the seven bonds, his total interest was

$$\text{total interest} = (7)(\$1937.50) = \$13{,}562.50.$$

Since Albert sold the bonds for $1002.50 each, his selling price was

$$\text{selling price} = (7)(\$1002.50) = \$7017.50.$$

Therefore, Albert's total income was

$$\text{total income} = \$7017.50 - \$7000.00 + \$13{,}562.50 = \$13{,}580.00.$$

Page 66

a. $R = \dfrac{Pr}{p} = \dfrac{(\$1000.00)(0.098)}{\$1085.00} \approx 0.090 = 9.0\%$

b. $R = \dfrac{Pr}{p} = \dfrac{(\$1000.00)(0.0915)}{\$1012.50} \approx 0.090 = 9.0\%$

c. $R = \dfrac{Pr}{p} = \dfrac{(\$1000.00)(0.07875)}{\$1028.75} \approx 0.077 = 7.7\%$

SOLUTIONS TO "TRY ONE" EXERCISES

Page 69

a. par value = $1000.00
annual percentage rate = 0.08375
maturity date = 2002
$N = 2002 - 1997 = 5$
price = $99\frac{1}{2}$ = $995.00

$$R^* \approx \frac{NPr + P - p}{N\left(\frac{P + p}{2}\right)}$$

$$= \frac{(5)(\$1000.00)(0.08375) + \$1000.00 - \$995.00}{5\left(\frac{\$1000.00 + \$995.00}{2}\right)}$$

$$\approx 0.085 = 8.5\%$$

b. par value = $1000.00
annual percentage rate = 0.08
maturity date = 2029
$N = 2029 - 1997 = 32$
price = 110 = $1100.00

$$R^* \approx \frac{NPr + P - p}{N\left(\frac{P + p}{2}\right)}$$

$$= \frac{(32)(\$1000.00)(0.08) + \$1000.00 - \$1100.00}{32\left(\frac{\$1000.00 + \$1100.00}{2}\right)}$$

$$\approx 0.073 = 7.3\%$$

Page 73

The current price is given as 101% of $1000.00, or $1010.00. The number of days from December 1 to February 3 is

$31 + 31 + 2 = 64$ days.

This means that the accrued interest for each bond is

$I = PrT = (\$1000.00)(0.07875)\left(\frac{64}{365}\right) \approx \$13.81.$

Adding $10.00 broker's commission, we have a total purchase price for each bond amounting to

total purchase price = $1010.00 + $13.81 + $10.00 = $1033.81.

Thus, Ronald's cost for the eight bonds was

$(8)(\$1033.81) = \$8270.48.$

Page 75

The number of days from March 10 to October 8 is as follows.

22	March
30	April
31	May
30	June
31	July
31	August
30	September
7	October
212	

At 6.5%, the interest for 212 days would amount to

$$I = PrT = (\$1000.00)(0.065)\left(\frac{212}{365}\right) \approx \$37.75 \text{ per bond.}$$

Thus, the net income on Andrea's investment is

	$7962.00	selling (ending) price:	(10)($1000.00)(0.79625)
−	7500.00	purchase (beginning) price:	(10)($1000.00)(0.75)
−	30.00	selling commission:	(10)($3.00)
−	30.00	purchase commission:	(10)($3.00)
+	377.50	interest:	(10)($37.75)
	$779.50	net income	

A net income of $779.50 on an investment of $7560.00 (purchase price plus commission) for 212 days corresponds to an annual effective yield Y of

$$Y = \frac{\$779.50}{(\$7560.00)\left(\frac{212}{365}\right)} \approx 0.178 = 17.8\%.$$

SOLUTIONS TO "TRY ONE" EXERCISES

SECTION 4

Page 90

a. The *NAV* of Riverside Capital Eqauity Fund at the close of the day on February 27, 1996 was $14.13.

b. The net change of the fund was –$0.01 from February 26 to February 27, 1996. Therefore, the previous day's closing price was

$$\text{previous day's closing price} = NAV - \text{net change}$$
$$= \$14.13 - (-\$0.01)$$
$$= \$14.13 + \$0.01$$
$$= \$14.14.$$

Page 92

The total value of Bob and Stacey's 200 shares on February 27, 1996 was

$$\text{total value} = (\text{number of shares})(\text{net asset value})$$
$$= (200)(\$18.30)$$
$$= \$3660.00.$$

From Table 9, the *NAV* of Evergreen Aggressive Growth A on February 27, 1997 was $20.05. The total value of Bob and Stacey's 200 shares on this date was

$$\text{total value} = (\text{number of shares})(\text{net asset value})$$
$$= (200)(\$20.05)$$
$$= \$4010.00.$$

The total value of the 200 shares increased from $3660.00 to $4010.00. The amount of the increase was

$$\text{amount of increase} = \$4010.00 - \$3660.00 = \$350.00.$$

Page 94

a. From Table 9, the one-year total return was 25.7%.

b. $\text{net income} = (\text{one-year total return})(P)(N)$
$$= (0.257)(\$16.12)(150)$$
$$\approx \$621.43$$

Page 95

From Table 9, the *NAV* of Fidelity Nordic on February 27, 1997 was $14.43. The purchase price for 250 shares was therefore $(250)(\$14.43) = \$3607.50$. The load on Fidelity Nordic was 3%, so the dollar amount of the load is

$$\text{load amount} = (0.03)(\$3607.50) \approx \$108.23.$$

Page 97

a. The one-year total return for Fidelity Nordic on February 27, 1996 was 36.6%.

b. Since the one-year total return was 36.6%, and the expense ratio was 2%, we have

gross return before expenses = 36.6% + 2% = 38.6%.

Answers to Odd-Numbered Exercises

SECTION 1

1. a. $d \approx \$1.97$ **b.** 4.4%

3. a. $d \approx \$2.14$ **b.** 7.3%

5. a. $d \approx \$0.51$ **b.** 0.9%

7.

Year	Earnings Per Share	Price-Earnings Ratio
1991	$e = \$3.71$	$\dfrac{p}{e} \approx 9.1$
1992	$e = \$7.02$	$\dfrac{p}{e} \approx 6.1$
1993	$e = \$6.22$	$\dfrac{p}{e} \approx 7.4$
1994	$e = \$3.16$	$\dfrac{p}{e} \approx 15.3$
1995	$e = \$2.53$	$\dfrac{p}{e} \approx 12.3$

9.

Year	Earnings Per Share	Price-Earnings Ratio
1991	$e = \$4.59$	$\dfrac{p}{e} \approx 8.9$
1992	$e = \$0.05$	$\dfrac{p}{e} \approx 875.0$
1993	$e = \$5.54$	$\dfrac{p}{e} \approx 10.0$
1994	$e = \$4.14$	$\dfrac{p}{e} \approx 13.1$
1995	$e = \$4.16$	$\dfrac{p}{e} \approx 15.1$

11. 1987:

Dividend Yield ≈ 0.8%

Annual Return ≈ 56.3%

1988:

Dividend Yield ≈ 1.2%

Annual Return ≈ 4.5%

1989:

Dividend Yield ≈ 2.1%

Annual Return ≈ −7.7%

1990:

Dividend Yield ≈ 2.6%

Annual Return ≈ −16.7%

1991:

Dividend Yield ≈ 2.4%

Annual Return ≈ −4.4%

1992:

Dividend Yield ≈ 1.1%

Annual Return ≈ 24.7%

1993:

Dividend Yield ≈ 0.7%

Annual Return ≈ 65.5%

13. 1987:

Dividend Yield ≈ 0.4%

Annual Return ≈ 44.4%

1988:

Dividend Yield ≈ 0.5%

Annual Return ≈ −8.1%

1989:

Dividend Yield ≈ 0.8%

Annual Return ≈ −11.5%

1990:

Dividend Yield ≈ 1.4%

Annual Return ≈ −30.6%

1991:

Dividend Yield ≈ 0.8%

Annual Return ≈ 79.6%

1992:

Dividend Yield ≈ 1.0%

Annual Return ≈ 24.1%

1993:

Dividend Yield ≈ 1.2%

Annual Return ≈ 14.2%

1994:

Dividend Yield ≈ 1.1%

Annual Return ≈ 29.6%

15. 1988:

Dividend Yield ≈ 3.9%

Annual Return ≈ 20.9%

1989:

Dividend Yield ≈ 6.1%

Annual Return ≈ −21.7%

1990:

Dividend Yield ≈ 9.5%

Annual Return ≈ −27.2%

1991:

Dividend Yield ≈ 6.5%

Annual Return ≈ −0.9%

1992:

Dividend Yield ≈ 1.9%

Annual Return ≈ 177.4%

1993:

Dividend Yield ≈ 1.1%

Annual Return ≈ 68.3%

1994:

Dividend Yield ≈ 1.8%

Annual Return ≈ −6.3%

1995:

Dividend Yield ≈ 3.3%

Annual Return ≈ 16.2%

SECTION 2

1. 80,200
3. a. $30,375.00
   b. $9575.00
   c. $9437.50
5. Previous day's closing price = 40
7. $18\frac{1}{8}$
9. $55.20
11. $131.25
13. Net Income = $3633.72
15. Net Income = $5462.90
17. Total Income = $10,825.00
19. Purchase Price = $14,750.00
    Selling Price = $138,000.00
21. Net Income = $20,075.00

SECTION 3

1. **a.** 30 years
   b. $5000.00
   c. Interest = $212.50
   d. Total Income at Maturity = $12,750.00
   e. Total Income = $375.00

3. **a.** $R \approx 10.4\%$
   b. Total Income = $12,375.00
   c. $R^* \approx 10.1\%$

5. **a.** 30 years
   b. $2981.25
   c. At Maturity = $9693.75
   d. Total Income = $1608.75

7. **a.** $R \approx 10.8\%$
   b. Total Income = $8085.00
   c. $R^* \approx 10.8\%$

9. **a.** $R \approx 10.6\%$
       $R^* \approx 10.6\%$
   b. $R \approx 7.9\%$
       $R^* \approx 8.5\%$
   c. $R \approx 8.7\%$
       $R^* \approx 10.1\%$
   d. $R \approx 9.9\%$
       $R^* \approx 10.0\%$

11. Annual Coupon Rate = 8%, Year of Maturity = 2002

13. **a.** 122   **b.** 119

15. **a.** $147.07   **b.** $4815.82

17. Net Income = $12,516.78
    Effective Yield $\approx 18.9\%$

SECTION 4

1. $18.31

3. Best: Robertson Stephens Contra Fund
   Worst: Riverside Capital Fixed Income Fund

5. a. 12.4%
   b. Net Income = $249.24

7. a. Purchase Price = $3668.50

   The load was $174.25.

   b. Total Cost = $3842.75

9. Fidelity Equity Income, Fidelity Equity Income II, Price Blue Chip, and Price International Stock.

SECTION 5

1. $d \approx \$1.91$

   Dividend Yield $\approx 4.5\%$

3. Earnings per Share = $e \approx \$8.35$

   P–E Ratio ≈ 3.9

5. Net Income = $5187.50

   Net Annual Return $\approx 34.6\%$

7. a. Annual Interest per Bond = $80.00

   Annual Interest for 20 bonds = $1600.00

   Total Interest for 20 Bonds = $12,800.00

   b. 32 years

   c. $R^* \approx 7.3\%$

9. $12.05